Extractive Reserves

IUCN - THE WORLD CONSERVATION UNION

Established in 1948, IUCN - the World Conservation Union - is an organization whose members include governments, non-governmental organizations (NGOs), research institutions and nature conservation agencies in 130 countries. IUCN's objectives are to promote and encourage the sustainable conservation of natural resources.

IUCN'S FOREST CONSERVATION PROGRAMME

IUCN's Forest Conservation Programme coordinates and supports the activities of the IUCN secretariat and members working with forest ecosystems, as well as research and promotion of the sustainable use of forest resources. The World Conservation Monitoring Centre (WCMC) supplies information on animal and plant species, and on habitats which are especially important for the conservation of biological diversity and the forest ecosystems.

The programme includes a study of forest policy, and field projects relating to specific problems arising with the management of the biologically most important forest resources. The principles of the World Conservation Strategy are applied in these projects, which combine the needs of conservation and those of local populations. Special emphasis is placed on setting up buffer zones around national parks and reserves.

IUCN's policy and activities are based on information supplied by its members or originating from field projects, and on the analysis of current trends prepared by WCMC. The programme is developed in consultation with international cooperation organizations, in order to ensure full consistency between development projects and conservation priorities.

IUCN publications contribute with information and technical recommendations to governments, international institutions, persons responsible for preparing development plans and conservation specialists. The Forest Conservation Programme's principal financial backer is the Government of the Netherlands.

COMMISSION OF THE EUROPEAN UNION

Twelve European countries (Belgium, Denmark, Spain, Germany, Greece, France, Ireland, Italy, Luxembourg, the Netherlands, Portugal and the United Kingdom) signed three treaties which provided the framework for the construction of a united Europe: one of them dealt with coal and steel, another with atomic energy, while the third and best known of the three set up the European Economic Community in 1958, with the purpose of joining together the economic resources of the 12 countries into a union, in which goods, persons, services and capital circulate freely, and within which trade, agriculture, transport and other economic sectors are governed by Community policy. There are five community institutions, including the Commission, which proposes and manages common policies, and monitors compliance with the treaties.

CNPT - CENTRO NACIONAL PARA O DESENVOLVIMENTO SUSTENTÁVEL DAS POPULAÇÕES TRADICIONAIS (National Centre for the Sustainable Development of Traditional Communities)

The CNPT is a unit which manages and coordinates plans, programmes and projects dealing with traditional communities, directly under the Presidency of the IBAMA, the Instituto Brasileiro do Meio Ambiente e dos Recursos Naturais Renováveis (Brazilian Institute for the Environment and Renewable Natural Resources).

The CNPT, whose activities are centrally coordinated in Brasilia, has started with six regional offices, set up in response to existing requirements in the towns of Rio Branco (AC), Porto Velho (RO), Macapá (AP), Manaus (AM), Imperatriz (MA) and Belém (PA). With regard to traditional communities, the main aim of the CNPT is to provide support for the preparation, implementation or introduction of projects, programmes and actions, either required by those communities or requested by them, directly through their representative bodies, or indirectly through governmental or non-governmental authorities set up for the purpose.

IUCN Forest Conservation Programme

Extractive Reserves

Editors
Julio Ruiz Murrieta and Rafael Pinzón Rueda

IUCN-World Conservation Union

Commission of the European Communities

Centro Nacional para o Desenvolvimento Sustentado
das Populações Tradicionais

1995

Published with the financial assistance of the Commission of the European Union.

Published by: IUCN, Gland, Switzerland, and Cambridge, UK

IUCN
The World Conservation Union

CNPT

ISBN: 2-8317-0273-9

Citation: Murrieta Ruiz, J. and Rueda Pinzón, M. Eds., (1995), *Extractive Reserves.* IUCN, Gland, Switzerland, and Cambridge, UK.

Translated by: Peter Whiting

Printed by: SADAG, Bellegarde-sur-Valserine, France

Cover design by: James Butler

Cover photograph: Rubber tapping. *Photo:* Miguel von Behr, CNPT

Available from: IUCN Publications Services Unit
219c Huntingdon Road, Cambridge CB3 ODL, UK

or

IUCN Communications Division
Rue Mauverney 28, CH-1196 Gland, Switzerland

The text of this book is printed on Reprise Matt and contains 80% recycled fibre.

Dedication

To the communities of the Extractive Reserves, whose living experience has made it possible to devise and implement the scheme, and especially to Chico Mendes, whose blood brought the idea to life.

CNPT TECHNICAL/ADMINISTRATIVE SUPPORT TEAM

CNPT / Brasília:

— Rafael Pinzón Rueda (Director of the CNPT)
— Adalberto Iannuzzi Alves (Techncial Advisor for the Amazon Region)
— Manoel Lima Feitosa (Technical Advisor for the Amazon Region)
— Carlos Aragón Castillo (Project Coordinator)
— Miguel von Behr (Techncial Advisor Atlantic Forest and Centre-South Region)
— Eneida Rodrigues da Silva (Administrative Advisor)
— Karla dos Santos Soares (Secretary)
— Walter Honorio de Lacerda (Computer)

CNPT / Rondônia:

— Maria de Lourdes Souza (Administrative Assistant)
— Melanias Vieira Neto (CNPT Representative)

CNPT / Acre:

— Marisanta Farias Nóbrega (CNPT Representative)

CNPT / Amapá:

— Wilson Menescal de Souza (CNPT Representative)
— António Sérgio Monteiro Filocreão (Technician)
— Mário Sérgio dos Santos Ribeiro (Technician)
— Mauro Jackson da Silva Moraes (Technician)
— Loudival Trinidade Romano (Administrative Assistant)

CNPT / Santa Catarina:

— Vitor Dutra (CNPT Representative)

CNPT / Maranhão:

— Edson dos Santos (CNPT Representative)

CNPT / Pará:

— Nelson Chada (CNPT Representative)
— Merlene Nayo (Technician)
— Itacelma Simões (Technician)

CONTENTS

PREFACE xi

INTRODUCTION (Rafael Pinzón Rueda) 1

HISTORICAL DEVELOPMENT OF EXTRACTIVISM (Rafael Pinzón Rueda) . 3

SOCIAL ORGANIZATION OF EXTRACTIVE COMMUNITIES
(Rafael Pinzón Rueda) 13

VIABILITY OF THE EXTRACTIVE RESERVES (Carlos Aragón Castillo) 19

GUIDELINE FOR THE CREATION AND LEGALIZATION OF
EXTRACTIVE RESERVES (Portaria IBAMA N° 51 de 11.05.94) 37

MANAGEMENT OF EXTRACTIVE RESERVES (Carlos Aragón Castillo) 49

EXTRACTIVE RESERVE OF ALTO JURUÁ (Manoel Lima Feitosa) 57

CHICO MENDES EXTRACTIVE RESERVE (Manoel Lima Feitosa) 69

MARINE EXTRACTIVE RESERVE OF PIRAJUBAÉ (Miguel von Behr) 77

FREXAL EXTRACTIVE RESERVE (Miguel von Behr) 81

RIO CAJARÍ EXTRACTIVE RESERVE (Adalberto Iannuzzi Alves) 85

RIO OURO PRETO EXTRACTIVE RESERVE (Adalberto Iannuzzi Alves) 97

EXTRACTIVE RESERVES OF CIRIACO, MATA GRANDE AND
EXTREME NORTH OF TOCATINS (Miguel von Behr) 105

MAPS 116

BIBLIOGRAPHY 129

PREFACE

Among IUCN projects, the study on moist tropical forests managed by traditional communities has attracted considerable attention . Through a number of studies undertaken as part of the Forest Conservation Programme, the IUCN has been trying to add to current knowledge of tropical moist forest ecosystems and of the interactions between these and traditional communities, with a view to informing the local population on topics related to conservation and the sustainable development of renewable natural resources.

The idea for this book emerged at a meeting on extractive reserves, which was held in Manaus, Brazil, in October 1992, and the project was jointly undertaken by the CNPT in Brazil and the IUCN.

The publication has been financed by the Commission of the European Communities and produced by the CNPT's technical team. It deals with the subject of Extractive Reserves, presented here as a model of forest management which succeeds in combining sustainable development and the conservation of natural resources.

Chapter I deals with the historical development of extractivism, from its origins starting with the first extractive activities harvesting rubber and nuts in Amazônia, through the phases of extraction of babaçu, hardwoods and minerals, the major social conflicts of the 1970s and the 1980s in the Brazilian Amazon region, unitl its adoption as an official concept embodied in national laws.

Chapter II describes the initial stages of social organization among the extractive communities, and analyses the first formally constituted bodies, such as the Fishing Colonies, the first Rural Workers' Unions, the consolidation phase, and lastly prospects for the future.

Chapter III looks at the viability of extractive reserves, concentrating on the extractive economy, the law regarding property and natural resources, criticism of the extractive reserve model, and ways of analysing the economic viability of the reserves. In addition, it presents a proposal for the sustainable development of extractive reserves, based on a diversification of subsistence production, on the handling of hydrobiological resources, on the improvement of traditional products and alternatives, on forestry and agricultural development, and on the rescue and support of other traditional activities. Lastly, a strategy is suggested to ensure the success of the above-mentioned proposal.

Chapter IV traces the steps required for the establishment and legalisation of extractive reserves in Brazil, from the first step, which is a formal request by local inhabitants for their area to be converted into a reserve, up to the preparation of the extractive reserve development plan.

Chapter V deals with the management of extractive reserves and analyses the laws which govern the reserves, the role of the CNPT and the importance of community organization.

Chapters VI to XII describe the existing nine extractive reserves in Brazil: Alto Juruá and Chico Mendes in the State of Acre; Ciriaco, Frexal and Mata Grande in the State of Maranhão; Rio Cajari in the State of Amapá; Rio Ouro Preto in the State of Rondônia; Extremo Norte de Tocantins in the State of Tocantins and Marinha do Pirajubaé in the State of Santa Catarina. The following aspects are described for each reserve:

geographical and environmental data, historical and socio-economic aspects, management and social organization, utilization plan, current projects and main problems.

The presentation of the different subjects considered in this book shows that the extractive reserve model is a real alternative not only for the tropical forests of Latin America, but also for the whole of the world's tropical moist forest ecosystem, including the continents of Africa and Asia.

Current trends show that this model can provide a basis for building a new type of society, which is respectful of local cultures and the environment and which achieves a fairer distribution of wealth.

Don Gilmour
IUCN Forest Conservation Programme

INTRODUCTION

Rafael Pinzón Rueda

Most of what has been written about extractive reserves has been as the result of studies by authors who have spent some time in the region and then have added to their knowledge through bibliographical research. This is not the case with this publication.

The extractive reserves report has been prepared by five technicians who have been experiencing life in the reserves on a day-to-day basis and are in charge of the CNPT/IBAMA programme (Centro Nacional para o Desenvolvimento Sustentado das Populações Tradicionais) .It is not a question of belittling the work of others who have thought and written about the subject, but of placing more emphasis on the experience of individuals who are actually living with the inhabitants of the reserves. Between them, the five technicians total 68 years of experience of working in the Amazon region with the extractive project. Two of them were actually born, bred and educated in the region (one in Acre and the other in Amazonas). The style is therefore rather that of authors who are on the spot and tell what is happening around them; it is spontaneous, reflecting what is to them a daily routine.

The tales of tourists and visitors to the Amazon of thirty years ago boil down to a sense of awe at that "huge green carpet", which was the impression gathered from an aircraft, and a feeling of emptiness, of loneliness, derived from the stereotyped image of "being lost in the forest". Nowadays, anyone flying over the region tends to be appalled at the way the "huge carpet" has been slashed and burned, particularly in the states of Pará, Maranhão, Mato Grosso, Rondônia and Acre.

The flight in a single-engined aircraft from Belém to Tucuruí, Carajás, Marabá or Imperatriz, or from Vilhena to Ariquemes, Porto Velho or Rio Branco fills visitors with dismay. No one can help noticing the devastation which has occurred in the forest. International reporters photograph apocalyptic, desolating scenes. The world is appalled. The northern countries are afraid of the "greenhouse" effect and they are all clamouring for the conservation of the environment, for the protection of the "huge, empty green carpet", the heritage of mankind, a pointer to the ecological equilibrium of the planet... but nobody ever talks about the "forest-dwellers", those who live inside the huge carpet. The extractive reserves are the vision of people who actually trod the ground, who stopped looking from a distance, who entered the forest and found a different reality, still ignored by those doing research on the Amazon: what needs to be preserved is the man who has dwelt there since before the Europeans discovered the New World, and having lived there for centuries has harmoniously preserved nature; and so it would have continued, had it not been for the ambitions of outsiders who came from other regions armed with tractors, chain-saws, fire and destruction, bringing with them the economic, ecological and social disaster of cattle-breeding, mining and timber depredation.

The one who needs saving is the real Amazonian, the "caboclo" and the "indio", the peasant and the Indian, the *"ribeirinho"* and the *"pescador"*, the river dweller and the fisherman, the "seringueiro", the rubber-tapper, and the *"castanheiro"*, who gathers the nuts, because only he can save the Amazon, if he continues to live there in harmony with nature, as a sentry against destruction, on the furthest lakes, rivers and waterways, the rubber plantations, the flood lands, the highlands and the river springs. The extractive reserves came with this mission.

It is a terrible mistake to describe the Amazon as a great demographic vacuum. It is far preferable to modernize the concepts of human geography and adapt them to ecology as the science of man's interaction with nature. It would then be found that the ecological equilibrium and sustainability of the Amazon depend on the spatial distribution of the population, according to the wise, traditional extractive approach, i.e. adapted to the spatial distribution of plant and animal species. From there, one moves on to a new concept of human occupation, whereby the "optimum" index is the one which secures sustainability. In the case of rubber-tappers, nut gatherers and fishermen, this works out at 5 km^2 per family (or 1 inhabitant per km^2). The conclusion then becomes that the Amazon is not a vacuum, but is in fact suitably occupied by about 4.5 million inhabitants in the rural area, with an average occupancy rate of 1.5 inhabitants per km^2.

The extractive reserves are not a panacea for Amazonia's problems, nor a magic solution for its occupation, if only because over half of the area has already been occupied and colonized under other, especially cattle-farming projects. Out of the total, maybe only 20% of the land fulfills the conditions required by law for extractive reserves. Precisely because the scheme is already late starting, every effort should be made to speed matters up, in order to protect whatever is still left to be protected.

This account of the extractive reserves provides one practical example of sustainability. A great deal of theory has been written on the subject, and countless meetings, workshops, courses and seminars have attempted to describe its content. The extractive reserve is the end materialization of all this. It not only provides an answer to the question of sustainability; it also embodies a rejection of monoculture, with today's extractive workers increasingly tending to diversify their economic activities.

It should also become apparent from reading this report that the extractive reserve scheme overcomes one of the most serious problems of developing countries, namely the question of the fair distribution of land, or agrarian reform, in that it fulfills the social purpose of the land by guaranteeing the legal rights of those who were already entitled in practice because they lived there traditionally.

The chief merit of the extractive reserve is that, unlike the other conservation schemes, it gives priority to people; man as conservation actor and manager catalyzes the whole process, in a genuine anthropogeographical sense. Just as there are the well recognized cultures of the Eskimos, the Tuaregs, the Polynesians and the wise African peoples, who preserve nature by making woods, rivers and animals sacred, so there is the culture of the "Amazonian peasant"; the reserves are another way of recognizing and rescuing that culture; and the joint management by the state and the communities is its practical expression.

One of its striking features is patience, forged in the concept of time as something static, like the immensity of the forest, the lakes, the slowness of the rivers and the interminable journeys. It was thanks to this patience that these people persevered in their patient search for new solutions some seventy years ago, in the 1920s, when rubber ceased to provide an easy answer. It was thanks to this patience that they stayed on in the forest, despite the disastrous, false solutions imported by exogenous cultures, based on destruction, like cattle-breeding, monoculture and mining. And it was thanks to this same patience that they were able to resist the expulsions and the deforestation and through their resistance found and proposed their own solution: extractive reserves.

Mythologies and religions have embodied our innate desire for happiness in the idea of a paradise, and this was conceived as a garden, a place with many trees, fruits, rivers and animals. Is materialism now destroying this product of collective conscience? What can we do to make sure that these corners of paradise continue to exist on our planet? These accounts of the reserve were written in order to encourage others to contribute, always in the hope of putting into practice the Roman law maxim *"salus populi suprema lex esto"*. Indeed the main purpose of setting up the extractive reserves must be the well-being of the people.

HISTORICAL DEVELOPMENT OF EXTRACTIVISM

Rafael Pinzón Rueda

Before considering the historical development of extractivism, it might be advisable to define the concept. The term "extractivism" is generally used to describe any gathering of natural products, whether of mineral (mining), animal (skins, meat, fats) or plant (woods, leaves, fruits, etc.) origin.

Some authors, such as Stephen Bunker (1985), take the concept of extraction to include forms of production through which a region is gradually impoverished, for the benefit of another region (this concept includes cattle-raising and agriculture).

Some say that "extraction" means any form of appropriation of resources, the natural occurrence of which has not been intentionally increased for purposes of exploitation. This definition would include the use of renewable and non-renewable resources, without taking into account the intensity of use, whether sustainable or not.

Since the aim of this publication is to portray the experience acquired with extractive reserves in Brazil, we shall understand extractivism as the sustainable use proposed for the reserves, that is, as the rational gathering in the biotope of renewable resources for market supply.

Brazil took its name from the extraction of *"pau-Brasil"* or brazilwood (Caesalpinia eclimata). Extractive activity has been a constant feature of the country's economic history. During the colonial period, it was mainly concentrated on wood and minerals, especially gold.

In the 19th century, extractive activities moved northwards, looking for woods, medicinal plants, cocoa and rubber. In this century, extraction in the Amazon region has been mainly related to rubber (Hevea brasiliensis) and the Brazil-nut (Bertolletia excelsa); after the second World War, there was an increase in the extraction of wood, which in the last five years has come under better control. Mineral extraction increased in the region after the 1960s, especially gold, iron, bauxite and cassiterite.

The *babaçu* palm (Orbygnia martiana) is extracted in some states of the north and north-east, and *carnaúba* (Opernicia cerifera) in the north-east. Woods, medicinal and ornamental plants, forest flowers and fruits are extracted locally in other regions of the country and a great many communities live on local fishing.

This outline of historical development is intended to facilitate an understanding of the origin of extractive reserves; it is not an analysis of extractive activity in Brazil as a whole, but only of the historical aspect related to the reserves. Thus specific attention will be given to the historical development of extractivism in the north of the country, where most of the reserve experiments have been taking place.

Extractivism in Amazonia

Economic interest in Amazonia was awakened in the 18th century with the search for the so-called "jungle drugs", including medicinal plants, oils, resins, cocoa, skins, fish and dried meat. Although at the time some farms, raising cattle and growing cocoa, coffee and cotton crops, were already established on the banks of the main rivers, these amounted to very little compared with extractive activities. The work of the Indians and peasants greatly contributed to the growth of extractivism, although the Indians in most cases were persecuted and obliged to work by the settlers. The participation of the negro in extractive activities in Amazonia was not significant.

The occupation of Amazonia was motivated by extractive activities, especially during the second half of the 19th century, when some 400 000 families came to settle in the area from the north-east in search of rubber, as a result of the rapid increase in production required by rising demand in the United States and Europe. This became the so-called "rubber cycle", which had its golden years around the turn of the century and declined after 1920.

In the second world war, there was a renewed interest in the extraction of rubber and thousands of north-eastern families were moved to the rubber plantations. After the war, the government tried to maintain a policy of incentives for the extraction of rubber, with credits for marketing and improvements. As the prices paid to producers were not attractive, however, rubber extraction went through a series of crises, with the result that in the last ten years a large number of families have abandoned the activity.

The extraction of rubber was always linked to that of nuts, which takes place in the same areas; rubber takes up the less rainy part of the year (May to November) and nuts the rainy period (December to March).

Characteristics of rubber and nut extraction

The extraction of rubber and nuts originated and developed on two main supports: outside capital and latifundia. These were two essential ingredients for the following reasons:

a) The undertaking required capital, along a chain running from the exporter who financed the town buyer, who in turn financed the middlemen and the latter the rubber planters, who provided sustenance for the tappers.

b) In order to be profitable, extraction needed large estates, owing to the low density of rubber and nut trees in the forest.

   The "traditional rubber plantation", a socio-economic feature of the region, had the following characteristics:

   — exclusively extractive activities and no agriculture;

   — imports of main consumer goods, leaving rubber workers completely dependent on the employer;

   — production cost guaranteed by the employer, but without leaving any profit margin for the rubber tapper; this used to be a way of preventing the latter from leaving the work, as he was always kept in a state of indebtedness to his employer;

   — excessive discounts in rubber weighing and low price fixing.

There are few rubber plantations nowadays which are still operating according to the traditional model. With the opening up of highways and branch roads and the advance of the agricultural frontier, many middlemen have appeared and little by little the figure of the single boss, the owner of the rubber plantation, or rubber planter, has been disappearing.

The traditional extractionist

The traditional extractionist in Amazonia inhabits the heart of the forest and lives by gathering natural resources, rubber and nuts, supplementing his income with hunting, fishing and gathering fruits such as *açaí, abacaba and patoá.*

Owing to the distances imposed by the dispersion of the productive species (three rubber trees or nut trees per hectare), he lives in isolation. His main characteristic is his total dependence on employers and the merchants who supply him with subsistence goods in exchange for his production.

As for his work, he has no employment tie with his employer, nor any lease contract; he does not own the land, but is entitled to practise his extractive activity, so long as he is willing to exchange his output for the goods supplied by the boss; in practice, he exchanges his labour for the supply of goods which are not available in the forest.

The extractionist adds to his income by selling a variety of farm products, but this income is drastically reduced in real terms owing to the extremely high prices he is obliged to pay for goods acquired from the traders. In some cases, the degree of isolation is such that there is no way of marketing the family's produce, so that the actual cash income is limited to what is obtained from the extractive activity.

Most of the extractive workers in Amazonia live in houses built on piles, with roofing and walls made of *paxiúba palm* leaves and/or straw; ducks, chickens and pigs live beneath the house, which leads to a proliferation of insects; above, the lack of protection facilitates the invasion of sickness-transmitting agents. The situation is further aggravated by the lack of hygiene. The most common illnesses are influenza, diahorrea, worms and malaria.

In the extractive areas, far from municipal centres, there are no schools, or if there are they operate in precarious conditions; hence the high illiteracy rate (70%), which is unlikely to fall, since the available schooling is not keeping pace with local population growth.

Practically all extractive workers in Amazonia do some fishing to supplement the food supply. Some families spend all their time fishing, on a manual basis. These people may be considered as extracting aquatic fauna and are not to be confused with industrial fishermen who come from other regions, with equipped boats and who fish with a view to marketing their produce in the large towns. Fish and wood resources are the most exploited in Amazonia, unfortunately by major economic groups.

Babaçu extraction

The *babaçu* palm, of the Arbignya and Ahalea varieties, is exploited mainly in the state of Maranhão, although it does grow in other states. Unfortunately, in the areas where it occurs most, cattle farms have appeared in the last 30 years, creating difficulties for the 400 000 or so families who derive most of their revenue from the extraction of *babaçu*. Before the arrival of the farms, *babaçu* gathering was a free and spontaneous activity. Now the cattle farmers, apart from pulling down most of the *babaçu*. groves, prohibit gathering in the remainder. There is an urgent need as a result to create conservation units, in order not only to protect the species, but also and mainly to allow the survival of the families.

The decline of extractivism in Amazonia

For 20 years, from 1965 to 1985, Brazil's policy was to occupy Amazonia, and to expand its agricultural frontier. The government opened up roads, distributed land, organized settlements and provided countless incentives, especially subsidised credits, to set up cattle farms and timber companies in the region. This policy of destroying the forest annihilated extractive activities in a number of regions, such as in the state of Pará, where the richest nut plantations of the country were pulled down, or in the state of Rondônia, where rubber and nut trees disappeared in the vicinity of the BR 364 highway. A number of extractive areas also disappeared in Mato Grosso, Maranhão and Acre.

This advance of the agricultural frontier into Amazonia destabilized extractive activities in several regions, transforming extractive areas into pastureland for cattle and extractive workers into agricultural labourers or small farmers.

As we shall see later on, the factors which halted this devastation were pressure from the actual inhabitants themselves, who exposed their lives to defend the forest, and the pressure brought to bear by international and national institutions. Even the government, from 1985 onwards, became aware of the importance of preserving the forests.

The action of timber companies in Amazonia

The extraction of woods in Amazonia, either for export or for consumption in the major domestic urban centres, was never a typical activity of forest dwellers. It was always an undertaking by major companies, through a front line of middlemen. These companies depend on government incentives, which have gradually been decreasing, and many of them disregard the law or abuse the terms of their concessions.

The genuine forest dwellers use woods privately for their own homes or for means of transport (such as small boats or canoes); the settlers, established there by the government, when pressed by the timber companies, start selling the timber at absurdly low prices. Settlers established by Incra in Pará, Rondônia and Acre are still now selling wood to the middlemen.

This pressure by the timber companies has greatly increased, and unfortunately research has shown that in the lowland regions traditional dwellers are also now selling woods to the middlemen (Whitesell, 1993).

Mineral extraction in Amazonia

Due to its considerable mineral wealth, both the government and the private sector have engaged in mineral extraction in Amazonia in the last 30 years. Major projects such as Carajás and Rio Trombetas in Pará extract iron and bauxite; Serra do Navio in Amapá extracts manganese, and Serra Pelada in Pará extracts gold; cassiterite is extracted in Porto Velho. In many rivers and indigenous areas, however, clandestine gold prospection is carried on, leading to one of the greatest environmental degradations in Amazonia, including water pollution; the government is constantly combating such activities.

The defence of extractivism in Amazonia: extractive reserves

From 1960 onwards, the governments of the northern states started a policy of attracting businessmen from the centre-south of the country to come and set up businesses in rural areas. In Acre, in the municipalities of Rio Branco, Sena Madureira, Plácido de Castro, Senador Guiomard, Xapurí and Brasiléia, many rubber plantations were transformed into pastureland for cattle, or abandoned. On the other hand, large plots of land where rubber tappers lived were acquired by the Federal Government for Agrarian Reform allotments.

In order to bring about these transformations, many rubber tappers were violently expelled from their homes and had to go and look for new areas. Many occupied abandoned rubber plantations or government land. These rubber tappers remained without employers and created the figure of the "independent rubber tapper", who is described as one who does not have a fixed employer to whom he is obliged to hand over his output because he owns the land. He is independent with respect to the old style of plantation owner, since the rubber tapper in fact continues to depend on tradesmen for supplies and for marketing his output. Although his independence is only very relative, it is still important, because from a social point of view, it enables him to take decisions freely and it marks a first step towards the conquest of a broader independence.

The independent rubber tappers of Rio Branco, Xapurí and Brasiléia were the first to come out in defence of extractivism, starting to organize in 1976 to prevent further deforestation of extractive areas. These movements, which came to be known as *"empates"* (stalemates), spread to other areas of Acre and even to other states (Amazonas and Pará).

The conflicts between farmers and rubber tappers impressed public opinion, especially after the death of the leader of the rural workers' union of Brasiléia, Wilson de Souza Pinheiro, in 1978. These same conflicts later were to lead to the assassination by farmers of the leader of the rural workers' union of Xapurí, Chico Mendes, in 1988.

In 1982, the Federal Government was beginning to consider the possibility of protecting extractive activities, when in Acre the *Instituto Nacional de Colonização e Reforma Agrária* (INCRA) and the *Superintendência da Borracha* (SUDHEVEA) ran into an impasse created by colonization projects, known as Managed Settlements; this was due to the fact that the rectangular plots set aside by INCRA for settlers upset SUDHEVEA's efforts to increase the rubber output of extractive workers, insofar as the rectangles cut across and broke up the *"colocação"* (allotment), a rubber tapper's production unit which followed the natural, haphazard scattering of the trees. This situation gave rise to sharp conflicts between neighbours.

In consultation with extractive workers and trade union leaders in Rio Branco, Xapurí and Brasiléia, it was agreed in the end that the most important consideration was not so much individual title rights to plots of land, but keeping up the productive capacity of the forest. It therefore appeared unnecessary to divide up the land into symmetrical, rectangular plots, but preferable instead to find a way of keeping the families in their "allotments", where they could carry on their extractive activity. The solution proposed by the two institutions was the *"Concessão Real de Uso"* (concession of use) issued by the rubber plantation to residents. To begin with, the projects of Boa Esperança in Sena Madureira and Santa Quitéria in Brasiléia were proposed on an experimental basis. Unfortunately, owing to a lack of administrative continuity in the two institutions, the proposal came to nothing. At least one positive point was to emerge, in that the existing arrangements under the two projects were frozen, so that many rubber workers were able to stay in their allotments.

The most important feature to remember in the recent historical development of extractive activities is that the *"empate"* or stalemate movement against deforestation was to serve as a basis for the political and social maturing of extractive workers, to the extent that the *"empates"* required organization and cohesion. These two forces were to emerge from the strengthening of the unions, where, in view of the necessity to maintain unity, the principle of the "collective concessions" came to be accepted and developed. It must be admitted that in 1980, perhaps under the influence of the land distribution policy pursued by the government, the extractive workers showed a preference for individual plots. Even today there is still a minority which has not fully accepted the idea of not having an individual title right to a plot of land.

In 1985, the rubber tappers, coming together in Brasilia for their first national meeting, requested the end of the colonization of rubber plantations and asked for the latter to be allocated to them on a concession basis, as a way of maintaining extractive activities. The request was also aimed at solving the land problem and protecting the forest against the threat of deforestation for the benefit of cattle-farming.

This meeting of the rubber workers constitutes the historic point at which the request to set up the "extractive reserves" became for the first time official. On that occasion, the idea was debated at length and mulled over by the participants, who came from several regions, and especially from Amazonia.

It is also worth noting that in 1985, the National Plan for Agrarian Reform was being launched in Brazil and the whole attention of society was concentrated on that subject. The proposal to set up the extractive reserves to that extent appeared as "agrarian reform" for the extractive workers, insofar as the reserves were a way of legalizing the possession and recognizing the land rights of those who had worked and lived there for many years. It may be said, therefore, that the extractive reserves as a project were historically inspired by the spirit of agrarian reform, that is, as a means of ensuring that the land fulfilled its social purpose.

The second major objective of the extractive reserves, namely the defence of the environment, was implicit in the first, considering that the land conquest was aimed at maintaining extractive activity, and that preserving the latter implied respect for the forest and its resources.

After that first meeting of the rubber tappers, INCRA again took up the problem. In its Directive No. 627 of 30 July 1987, it then proposed as a solution setting up the *Projeto de Assentamento Extrativista* or PAE (Extractivist Settlement Project), "for the purpose of exploiting areas containing extractive rubber plantations by means of economically viable and ecologically sustainable activities, to be pursued by the communities who occupy or might come to occupy the aforementioned areas". The Directive established that activity in the area was to be "subject to a concession of use in a communal system, in the form decided by the community of beneficiaries - in association, co-ownership or cooperative".

INCRA's official decree meant the incorporation of the extractive reserves (under the designation of Extractivist Settlement Projects, PAE) as part of the National Plan for Agrarian Reform. By April 1994, ten extractivist settlement projects had been created, five in Acre, three in Amapá and two in Amazonas, totalling 889 548 hectares. Only three of these, in Acre, received support at the start.

Extractivist settlements

PROJECT	STATE	AREA (ha)	FAMILIES
Sãao Luiz Remanso	Acre	39,572	130
Cachoeira	Acre	24,973	80
Santa Quitéria	Acre	44,000	150
Porto Dias	Acre	22,145	83
Riozinho	Acre	35,896	120
Maracá I	Amapá	75,000	214
Maracá II	Amapá	22,500	94
Maracá III	Amapá	226,000	760
Antimary	Amazonas	260,277	867
Terruaã	Amazonas	139,235	426
TOTAL		**889,598**	**2,924**

Source: INCRA, 1989

From the beginning, social movements joined in the struggle to defend extractivism and to look for solutions, initially to the land problem. Gradually demands were channelled through the *Conselho Nacional dos Seringueiros,* the CNS (National Council of Rubber Tappers), set up in 1985. In view of the lack of progress achieved by the Extractivist Settlement Projects (PAE) and of growing domestic and international pressure in favour of the preservation of the forests, the CNS, apart from continuing to advocate the creation of extractive reserves as a means of implementing agrarian reform for extractive workers, started to defend them as "territorial areas for sustainable use and the conservation of renewable natural resources". This was a new victory, insofar as the extractive workers began to understand and defend the ecological dimension of the scheme. In other words, they were beginning to grasp the theory behind the "praxis".

The Federal Government also took a step forward, by legalizing the creation of extractive reserves as part of the national environmental policy in Law No. 7.804 of 18 July 1989, to which was added Decree No. 98.897, of 30 January 1990. According to those laws, the institution responsible for the reserves became the IBAMA, *Instituto Brasileiro do Meio Ambiente e dos Recursos Naturais* (Brazilian Institute for the Environment and Natural Resouces). According to IBAMA's Directive No.22-N of 10 February 1992, the management body responsible for reserve matters is the CNPT, *Centro Nacional para o Desenvolvimento Sustentado das Populações Tradicionais* (National Centre for the Sustained Development of Traditional Communities).

By April 1994, nine extractive reserves had been set up, according to the following table:

Extractive reserves

NAME	State	MUNICIPALITY	DECREE OF ESTABLISHMENT	AREA (ha)	POP. (inh)	MAIN ACTIVITY
PIRAJUBAÉ	SC	Florianópolis	533/92-20.05	1,444	690	*Berbigão* shellfish
CIRIACO	MA	Imperatriz	534/92-20.05	7,050	844	*Babaçu* and subsistence agriculture
EXT. NORTE DO TOCANTINS	TO	Augustinópolis, Sampaio	535/92-20.05	9,280	320	*Babaçu* and subsistence agriculture
QUILOMBO DO FREXAL	MA	Imperatriz	536/92-20.05	9,542	1,080	*Babaçu,* subsistence agriculture and fishing
MATA GRANDE	MA	Imperatriz, João Lisboa	532/92-20.05	10,450	776	*Babaçu* and subsistence agriculture
ALTO JURUÁ	AC	Cruzeiro do Sul, Taumaturgo de Azevedo	98.863/90 -23.01	506,186	5,821	Rubber
CHICO MENDES	AC	Rio Branco, Xapuri, Brasilea, Sena Madureira, Assis Brasil	99.144/90 -12.03	970,570	12,017	Nuts, *copaiba* and rubber
RIO CAJARI	AP	Laranjal do Jarí, Masagão	99.145/90 -12.03	481,650	3,639	Nuts, *copaiba,* rubber and *açai*
RIO OURO PRETO	RO	Guajara Mirim	99.166/90 -12.03	204,583	775	Nuts, *copaiba* and rubber
TOTALS:				2,200,750	25,962	

Source: IBAMA/CNPT, 1994

Official conception of extractive reserves

According to Decree No. 98.897, the official conception of extractive reserves is as follows:

Art. 1. Extractive reserves are territorial areas designed for self-sustaining exploitation and the conservation of renewable natural resources, by extractive communities.

Art. 2. The Executive Power shall set up extractive reserves in territorial areas considered to be of ecological and social interest.

Single paragraph - Territorial areas shall be deemed to be of ecological and social interest if they possess natural or exemplary characteristics of the biotope such as to allow their self-sustaining exploitation, without jeopardizing environmental conservation.

Art. 3. The deed of establishment shall specify the geographical limits, the beneficiary population and the measures to be taken by the Executive Power for the establishment of the reserves, any necessary expropriations being the responsibility of the Instituto Brasileiro do Meio Ambiente e dos Recursos Naturais Renováveis - IBAMA.

Art. 4. Self-sustainable exploitation and the conservation of natural resources shall be regulated by contract of real concession of use, as specified in Article 7 of Decree-Law No. 271 of 28 February 1967.

Subparagraph one - The real right of use shall be conceded free of charge.
Subparagraph two - The contract of concession shall include the utilization plan approved by IBAMA and shall contain a termination clause in the event of any damage to the environment or the transfer of the concession "inter vivos".

Art. 5. IBAMA shall be responsible for supervising the extractive reserves and for monitoring the fulfillment of the conditions stipulated in the contract referred to above.

It is easy to see that, according to that decree, the extractive reserves cannot provide a development model for the whole of Brazil. They are only valid for some areas, where the conditions mentioned in the decree apply, namely:

1) Areas containing renewable natural resources and extractive communities.
2) Areas considered to be of ecological and social interest.
3) Areas guaranteeing self-sustainability subject to approval by IBAMA of a utilization plan.

In the light of events in existing reserves, of discussions with reserve dwellers, of the analysis of their economic activities, and of the proposals and aspirations they have expressed, the concepts have since evolved, giving rise to the following principles of understanding regarding the reserves:

— Extractive activities are not the only economic activities of the reserve, even though they may at present provide the basic sustenance of residents;
— Improvement in residents' living conditions must be sought by expanding and improving existing extractive and farming activities and by introducing new activities with no environmental ill effects;
— Improvements needed include:
  - increasing the output and productivity of existing forest products;
  - improving the marketing system;
  - adding value to the products, by encouraging local processing;
  - increasing farming activities, by making use of existing cleared land.
— Changes should be based on an associative approach, with joint management of the reserves;

— This associative approach must find ways and means of becoming independent with regard to supplies and marketing.

Conceptual evolution of extractivism

Until the beginning of the 19th century, as far as extractivism was concerned, the world was dominated by naturalist ideas, swept along by the success of botany, zoology, scientific discoveries and the major expeditions in Africa, Asia and Latin America. People then used to talk about "mother nature" and the immeasurable riches it contained.

With advent of the industrial revolution and especially with the influence of Marx's historical materialism, according to which everything depended on the economic order, the driving force behind all events, natural resources came to be known as "raw materials", seen as indispensable for satisfying a mad drive to transform them into new products to meet consumption and export targets.

At that time, raw materials were taken to be inexhaustible and their consumption controllable by man.

One century later, with technological advances, population growth and the excessive consumption of "raw materials", man began to change his views of extractivism. It was first observed that natural resources were not inexhaustible, and that they needed to be reproduced if they were to be preserved and even passed on to other generations. That gave rise to the idea of sustainability and practicing sustainable development. Extractivism became part of the new concept.

In the particular case of Brazil, where the defence of extractive natural resources emerged as a result of the struggle by extractive workers to defend their land, the goal pursued was not just sustainable, but "socially just" development. The extractive reserves must remain a means of putting justice into practice, by attributing land to those who have traditionally dwelt on it and defended it.

The conceptual development of extractivism in Brazil, through the direct participation of extractive workers, succeeded in taking that significant step forward, thereby establishing extractive reserves not only as an ecological victory, but even more so as a social victory.

We may consider extractive reserves as one milestone which has been reached, in the historical development of extractivism, since they embody several ideals pursued by contemporary society:

— A balance between development, environmental conservation and social justice;
— Participation by society as an agent and not an object in the process, considering that the reserves are self-managed by residents;
— Retrieval and refinement of popular knowledge, since the utilization plan of the reserves is based on the experience and wisdom of dwellers who have coexisted there for many years in harmony with nature;
— Lower cost of protecting forests, once the local residents have taken up their defense.

In the development of extractivism, the creation of extractive reserves is only one stage. A further effort is needed to implement the reserves and ensure that they are a success. The most urgent measures include:

— Strengthening local organizations, especially by training human resources;
— Supplying the material conditions and tools needed to implement the cooperative management of resources and extractive activity;
— Distribution of power over natural resources (Concession of use and Utilization plan);

— Release from ties of dependency on middlemen (supply of goods, inputs and marketing of production);
— Adding value to extracted products (enhancement or pre-industrialization)

The extraction of Amazonian timber for export was always controlled by major companies *(photograph: Julio Ruiz Murrieta).*

Medicinal plants, flowers and fruit are extracted from the forests in other regions of Brazil *(photograph: Julio Ruiz Murrieta).*

SOCIAL ORGANIZATION OF EXTRACTIVE COMMUNITIES

Rafael Pinzón Rueda

Background

The social organization of the extractive communities as it is nowadays can only be understood if the question is considered from an historical point of view. Profound changes have occurred in the last 20 years, running against an age-old logic which condemned extractive workers to isolation and disorganization.

Until a few years ago, it seemed hopeless to try to organize extractive workers in Amazonia, owing to their main characteristics, namely isolation in the forest, lack of communication and practically total independence for their subsistence. This kind of situation would discourage any theoretician of social organization. And yet, a more attentive look will show that even in the heart of the forest the Aristotelian definition of man as a "social animal" still holds. Even in the furthest removed localities of the region, ever since the very beginnings of occupation, there was a deep-seated notion of the *"seringal"* (stand or plantation of rubber trees) as the social unit and along the river banks the *"povoado"* (village) was always the embodiment of this unit. Little by little, the notions of the rubber plantation and the village were replaced by that of the "community", which was understood more in the geographic sense of families living in a particular place; in many areas the "community" is an embryo of the village, which in turn generates a large village and eventually the future small township. The "community" generally gathers for *"mutirão"*, which is a voluntary activity, such as work, which the group does together.

It should be recognized that the concept of "community" has also evolved and that this too is now being taken to mean an organizational unit of society. In many extractive areas, the community is understood to mean a group of families, which apart from living in the same physical space, organize to achieve common objectives, generally aimed at improving their living conditions.

This organizational function of the community was considerably fostered by the Catholic church, with the creation of "base communities", a feature introduced in the region in the last 30 years.

While the "social animal" aspect of the extractive worker has expressed itself through the rubber plantation, the village or the community, these manifestations cannot be considered as real forms of "social organization", in the strict sense, since the power to order society and to lay down rules was always held by the local boss. Social expressions are restricted to religion, leisure at times of festivities, or solidarity in difficult moments. Still now the bosses are opposed to any attempts at organization among the extractive workers.

Formal organizations

The oldest organizations of extractive workers in the region are the fishing colonies, which have operated since the 1950s, but which unfortunately did not pursue the question of organization with the membership.

The real social organization of the extractive communities originated with the trade union movement. The first rural workers' unions were founded in the region in the 1970s. Initially these unions were joined by small farmers and settlers set up by INCRA under settlement projects.

Extractive workers, who had always lived independently and who, as a result of the existing system of land ownership, had been accustomed not to feel the need for property titles over their allotments, equally did not feel the need to join any union. The significance of this fact was brought home when the agricultural frontier, with its deforestation, reached the extractive areas and began to drive them away and burn down their houses. The union then served as a catalyst in a difficult situation and united extractive workers who tried to defend their rights. A further task of the union was to dealing with pensions.

The rural workers' unions grew in size and importance as the agricultural frontier expanded and land disputes became more frequent in the states of Maranhão, Pará, Amapá, Amazonas, Mato Grosso, Rondônia and Acre.

The main merit of the unions with regard to the social organization of the extractive workers was that in the very heart of the forest, in the communities, they set up "union delegacies", which were component cells of the union. This inside approach helped the extractive workers to lose their fear of the union and to join. These delegacies serve as the focal point for countless meetings, courses and discussions, which have gradually been changing the mentality of extractive workers, making them more aware of the importance of their social organization. The delegacies and the unions have acted as real schools for the socialization of extractive workers and have provided the first cells of their organization.

From the 1970s onwards, another type of social organization appeared in the region, the "producers' associations", which in general were aimed at improving production and marketing. Some of these associations brought in extractive workers, as was the case with the 63 rubber tappers' associations (ASSERS), which were set up between 1982 and 1984, in the municipalities of Sena Madureira, Rio Branco, Plácido de Castro, Xapurí, Brasiléia and Assis Brasil, in Acre. The purpose of these associations was to ensure the independent supply of goods of prime necessity and the production and marketing of prime quality rubber: the Brazilian smoked leaf.

The idea of association among extractive workers has prospered. Nowadays, practically all the reserves have set up residents' associations. In some cases, within the same municipality, as in Brasiléia (AC), several producers' associations have combined into a group of associations.

Unfortunately, this type of rural and urban social organization in the form of associations also tends to be used at present in the Amazon region, especially by politicians, for purposes other than merely creating greater awareness, participation and solidarity in society. One might even go so far as to say that the basic idea has been distorted and that the associations have proliferated unduly.

Stronger organization

The social organization of the extractive populations has been strengthened through the work of the National Council of Rubber Tappers (CNS). The basis of this strengthening process has been the efforts of the unions.

As we mentioned earlier, the advance of the agricultural frontier, the destruction of the forest and the expulsion of the families living there forced the extractive workers to set up a system of defence. With the trade unions of Xapurí and Brasiléia, in Acre, providing the coordination, they set up operation "stalemate" *("empate")*, which consisted in families, adults and children, peacefully occupying places earmarked for destruction in order to prevent it. This type of operation, which spread throughout the north from 1978 to 1988, was staged over 40 times in Acre. It has to be well organized, since women and children are involved for days at a time; it was also through this type of exercise that union members gained a higher sense of cohesion and awareness, thus establishing a solid base for strengthening the social organization of extractive workers.

Chico Mendes led the *"empates"* and gave the movement its martyr figure, as he was assassinated by farmers who were afraid of the movement's strength. Before Chico Mendes, in 1978, the president of the Rural Workers' Union of Brasiléia, Wilson de Souza Pinheiro, had already been assassinated, when he was only peacefully defending the rights of the rubber tappers to remain on their allotments.

In 1985, the extractive workers held the first national meeting of rubber tappers, on which occasion they set up the CNS, Conselho Nacional dos Seringueiros (National Council of Rubber Tappers).

This council originated with the objective of "fighting for a system protecting the life of the rubber tapper on his allotment, and ensuring that he is not expelled from his land". Its first undertaking was to protect the lives of thousands of families living in the forest.

The CNS was set up to oppose the expropriation of land and the devastation of the forest; in all these years, it has been working extremely hard to expand the community organizations and to encourage the establishment of cooperatives.

The CNS's action in favour of stronger social organization has been decisive from the point of view of the prestige and the influence it gained both nationally and internationally. With CNS backing, for instance, the extractive reserve associations have already signed agreements in support of their activities with the following institutions:

- Banco Nacional de Desenvolvimento Econômico e Social BNDES
- Gaia Foundation
- McArthur Foundation
- Health Unlimited
- Universities of São Paulo and Campinas - USP - UNICAMP
- IDB - Inter-American Development Bank
- Swiss Embassy
- IDRC - Canada
- Oxfam
- Miserior
- Konrad Adenauer Foundation
- World Wildlife Fund - WWF
- Instituto Brasileiro do Meio Ambiente e dos Recursos Naturais Renováveis - IBAMA
- Austrian Embassy.

More recently, the CNS has concentrated its efforts on defending the prices and markets for extractive products (rubber, nuts, *açaí, cupuaçu, babaçu*), and to obtain marketing credit lines.

The degree of maturity achieved with the social organization of extractive communities is shown by two important features:

a) The acceptance, by the majority of the inhabitants of extractive reserves, of collective land use. This acceptance has emerged after many years and a great deal of discussion, clarifications and reflection. Influenced by the settlement system, which was based on the distribution of individual plots of land, to begin with the extractive workers did not easily come to terms with the idea of using the land collectively.

b) An understanding of the various roles fulfilled by the trade union and associations. Social organization requires specialization and a distribution of tasks. For someone who has focused his struggle for survival on the union and has been defended by the latter, it is very difficult to understand the role of an association within the same movement. Fortunately, the extractive workers gradually began to distinguish between the different roles and to give their support to the associations as well.

It is now becoming clear to all concerned that the union fulfils a political purpose, that of defending a class (of extractive workers), and providing education in the rights of free citizenship. The association, on the other hand, deals more with the aspects of improving production and marketing, as means of improving the standard of living.

Thanks to the progress achieved in the extractive associations themselves, special sections are now being set up to deal with production or with social matters, such as housing, education and health.

There is also the feeling that in the not to far distant future the associations may evolve towards a system of cooperatives, this time set up from the ground upwards, starting from concrete, efficient associative bases. This would be one way of salvaging the image of the real cooperative spirit. Unfortunately, cooperatives have been set up in the region in the past without any previous associative experience; all failed, leaving doubts regarding the validity of the cooperative system.

The social organization of extractive workers also achieved significant advances among fishermen and the Indians. The former have been organizing into more dynamic associations than the fishermens' colonies, including the MONAPE, the National Fishermen's Movement. The Indians have a number of organizations such as the CAPOIB (Coordenação das Articulações dos Povos Indígenas do Brasil), COIAB (Coordenação das Organizações Indígenas da Amazônia Brasileira), CTI (Centro de Trabalho Indigenista) or the UNI (União das Nações Indígenas).

The social organization of extractive communities may be described nowadays as having as its base the associations or unions, and on a higher level the National Council of Rubber Tappers (CNS), which channels aspirations generally. The CNS is part of the GTA, Groupo de Trabalho da Amazônia (Amazonian Work Group), an NGO which groups together over 250 NGOs and which deals with the government and with other national and international institutions. The GTA has representations in individual states dealing with other organizations.

Thanks to the strength of their organization, the extractive workers have managed to establish a bridgehead with the Brazilian Government, facilitating a dialogue and supporting their claims. It was the organizations of extractive workers which requested the establishment of the CNPT (Centro Nacional para o Desenvolvimento Sustentado das Populações Tradicionais, or National Centre for the Sustained Development of Traditional Populations) within the IBAMA (Instituto Brasileiro do Meio Ambiente e dos Recursos Naturais Renováveis, or Brazilian Institute for the Environment and Renewable Natural Resources).

Future prospects

It may be seen from the above study that the extractive workers are now at the first stage of their social organization, which consists in their awareness of the importance of the organization and in their introduction to institutional associative activity, in the form of associations and trade unions. A second stage is about to begin and its main objectives will be:

a) training organization leaders to manage their units;
b) training organization members to ensure a better distribution of tasks, as a means of avoiding the concentration of power and decision-making in a few hands;

c) training organization members to prepare, negotiate and implement projects of an economic nature. The practical financial results achieved will be the most convincing arguments for maintaining the organization.

There are two reasons for believing that the organization of extractive workers has a promising future:

a) the fact that these organizational movements have emerged from the struggle of extractive workers to defend their rights;
b) the survival of "extractive" families basically depends on continuing the struggle, and the successful continuation of the struggle depends on the ability to organize.

A traditional extractive family in Amazonia *(photograph: Julio Ruiz Murrieta).*

Extractive workers have been forced by the advance of the agricultural frontier, the destruction of the forest and the expulsion of families to organize their own defence *(photograph: Julio Ruiz Murrieta).*

VIABILITY OF THE EXTRACTIVE RESERVES

Carlos Aragón Castillo

The extractive economy

Recent studies have confirmed that communities living in the extractive reserves, in order to guarantee their survival and to reproduce as such, are obliged to develop different types of activities and consequently different products, of which part is used for self-consumption and the other part is marketed. In this process, the communities organize into family production units, in which the workforce of the members is distributed over the different activities, either of subsistence agriculture or properly speaking extractive, according to a typical extractive calendar (Table 01).

In the production units, known as *"colocações"* (allotments), some farming activity takes place on cleared patches of land and in back yards, where some fruit trees and vegetables may be grown and a few small animals reared, such as ducks, chickens and pigs. Extractive activities are carried on in the forest (hunting and gathering), as well as in the rivers and lakes, where the inhabitants fish.

The farmed areas are not in one piece, as is the case with traditional agricultural units. The land and the forest, for the purposes of agricultural production, are occupied discontinuously, with units of productive areas distributed at random, in accordance with the conditions imposed by nature in its physical and geographical aspects (proximity of rivers and waterways) or its biological characteristics (distribution of plant species).

In this way, the family production unit may be understood as a form of interaction between the workforce used in productive activities and respect for nature, implemented and facilitated by the application of a broad range of skills and knowledge accumulated in the course of several generations, which have survived by using the resources of the forest and have permitted their conservation.

Extractive activity implies a very close relationship between man and nature, one characteristic being its adaptation to natural cycles and conditions: " In other words, it is not a matter, as in other types of agricultural production, of adjusting the conditions of nature to those imposed by man" (Teixeira, 1980).

It is important to emphasize that extractive production is not restricted to a single product, but is oriented towards a multiple use of the forest, as may be observed in the Chico Mendes Reserve, which produces palmito, jatobá, *bacuri, cocoa, copaiba,* honey, *bacaba, patauá* and *açai* (CNS, 1994).

In that reserve, extractive activities contribute 62% of family income, agriculture 29% and livestock farming 9%. These percentages tend to vary in other reserves such as Alto Juruá, Rio Ouro Preto and Rio Cajarí, depending on regional characteristics.

Property rights and extractive resources

Forest resources are frequently subject to over-exploitation pressure. In some cases, the rapid growth of demand leads to a real increase in the prices of products, and hence to an increase in production, while in other cases, the population which depends on a resource grows at a rate exceeding the resource's regenerative capacity.

Table 01
Yearly calendar of extractive activities in the Rio Cajarí reserve

	ACTIVITIES	JAN	FEB	MAR	APR	MAY	JUN	JUL	AUG	SEP	OCT	NOV	DEC
01	CLEARING								X	X			
02	FELLING									X	X		
03	BURNING										X		
04	FINAL CLEARANCE											X	X
05	MANIOC PLANTING	X											X
06	MAIZE PLANTING	X											X
07	RICE PLANTING	X	X										
08	BEAN PLANTING					X							
09	FRUIT TREE PLANTING	X	X	X									
10	MANIOC HOEING							X	X				
11	MAIZE HOEING		X	X									
12	BEAN HOEING						X	X					
13	FRUIT TREE HOEING	X	X	X	X	X	X	X	X	X	X	X	X
14	MANIOC GATHERING	X	X	X	X	X	X	X	X	X	X	X	X
15	MAIZE GATHERING							X	X				
16	RIZE GATHERING							X	X				
17	BEAN GATHERING							X	X				
18	FRUIT GATHERING	X	X	X	X	X	X	X	X	X	X	X	X
19	GRAIN PROCESSING							X	X				
20	FLOUR MANUFACTURE	X	X	X	X	X	X	X	X	X	X	X	X
21	SALE OF PRODUCTION	X	X	X	X	X	X	X	X	X	X	X	X
22	ROAD CLEARING	X	X	X									X
23	SHED BUILDING	X	X										X
24	STORE BUILDING	X	X	X	X								
25	BASKETWORK	X	X										X
26	NUT GATHERING	X	X	X	X	X	X						
27	NUT BREAKING	X	X	X	X	X	X						
28	TRANSPORT TO STORES		X	X	X	X	X						
29	TRANSPORT TO MARKET	X	X	X	X	X	X						
30	NUT SALES	X	X	X	X	X							
31	AÇAI GATHERING	X	X	X	X	X	X	X	X				X
32	PALMITO EXTRACTION	X	X	X	X	X	X	X	X	X	X		
33	HUNTING	X	X	X	X	X	X	X	X	X	X	X	X
34	FISHING	X	X	X	X	X	X	X	X	X	X	X	X

Source: World Bank - IBAMA/CNPT - CNS, Socio-economic survey of the Rio Cajarí Reserve, 1993.

According to an analysis of market potential for extractive resources, under conditions of open access, an unrestricted increase in the extraction of native plant products could exhaust stocks and perhaps destroy them altogether. Pressures on resources may arise as a result of a boom in demand or a change of ownership rights.

Innovation in the agricultural industry or the introduction of synthetic processes may eventually meet the greater demand, while at the same time maintaining an ecological balance, but its probable effect will be to expel rural labour employed in extractive activities.

The initial distribution of property rights affects the distribution of profits derived from innovation. One modern economic theory of natural resources attributes the cause of resource depletion to insufficient individual control over their use, exercised through property rights. Privatization is the usual recipe of neo-classical theory, in cases where such depletion occurs.

Unlike what is suggested by the theories, however, a move towards preventing access to the resources, by concentrating them in fewer hands, often leads to an increase, and not a reduction, in the rate of exploitation (Repetto and Holmes, 1983).

This delimitation of property rights through privatization can also result in demographic pressures on the remaining, limited, communally managed resources, thus speeding up their depletion. Such results may also be termed the "tragedy of non-shared resources".

Property can withdraw the land factor from production (Taylor, 1978). This can result in growing pressure on the remaining available land, managed sustainably under a traditional system of common property. The "tragedy of non-shared resources" occurs when the resources are privatized in such a way that the benefits obtained by common management are eliminated.

According to liberal-style criticism, opposed to the former theory, the reserves will be unable to cope with the pressures of dwellers on the resources, precisely on account of the absence of private individual rights. According to this type of criticism, only the institution of private property rights (in other words, of large land-holdings over the former rubber plantations), would prevent the rubber tappers from falling victim to a "tragedy of shared resources" (Hardim, 1968; Dawes, 1973).

Although this argument has not been used directly by critics of the reserves, some specialists have taken the view that it deserves consideration. It may be argued against it that there are communal property systems in many parts of the world which exploit natural resources sustainably, so long as there are local organizations taking collective action (see Ostrom, 1990; McCay and Acheson, 1990; Berker, 1989).

The negative effects of selfish action by individuals on "communal areas" can be countered by appropriate rules and values and by the introduction of mechanisms for monitoring and repressing infringements, as provided by the management facilities of the extractive reserves.

Criticism of the extractive reserve model

In recent years, the viability of extractivist reserves has been hotly debated and they have been subjected to a great deal of criticism. This is mainly of two kinds: criticism on ecological grounds and criticism on economic grounds.

Criticism on ecological grounds

Among the ecological critics, some biologists and conservationists argue that the existence of human populations is inevitably prejudicial to environmental conservation and to preservation of biodiversity. In the case of the rubber tappers living in the extractive reserves of Amazonia, the critics maintain that these groups produce a predatory impact on wild animals and therefore on biodiversity as a whole (Oliveira, 1991; Peres, 1990).

Recent studies indicate that subsistence hunting in Amazonia, apart from contributing significantly to the food intake of extractive communities, affects the biodiversity of the forest, drastically reducing the populations of some of the species hunted. These same studies indicate that the negative effects of hunting can be controlled in three ways: by establishing a management plan, by introducing environmental education as a means of understanding such management and by implementing alternative sources of protein (Martins, 1992).

On the other hand, there are biologists who argue that overall biodiversity (as opposed to the abundance of some of the larger species) can be extraordinarily high in areas inhabited by rubber tappers, as was empirically observed in the area of the Alto Juruá reserve (Brown, 1981, 1990; Conservation International, 1991; Almeida, 1993). Other biologists maintain that biodiversity can increase in secondary forests (Stallings and Redford, 1991), provided that the impact of hunting is controlled through institutional mechanisms (Bodmer, Fang and Ibañez, 1988).

Criticism on economic grounds

Criticism of this kind has been based chiefly on the poor contribution of extractive reserves to rural welfare and to the social relations on the basis of which products are collected and marketed, which historically have been associated with situations of exploitation, such as debt peonage or manipulation by intermediaries.

In many cases, the gathering and rudimentary processing are done by people who have no control either over the resource or over the marketing of the product. These people therefore have little bargaining power to determine what proportion of the product's value they can retain; they are obliged to live in a state of dependency on landowners, concession holders or tradesmen, and are the first to suffer the consequences of the boom and bust cycles, which are endemic in extractivist activities.

Where demand is intense and the plant resource base is accessible but limited, extraction can lead to the rapid degradation of the resource. By limiting supply, the degradation will push prices up even more, which in turn will increase pressure on the resource. A theory explaining this process of market development, which leads to the degradation of extractive resources, was proposed by Homma (1981) and Repetto and Holmes (1983) as applicable to subsistence resources in general.

It has also been argued that, considering that the supply of the extractive product is inelastic, there is a tendency for the price of the product to increase, which will intensify domestication as an alternative to increase the supply, making it elastic and thereby lowering the price of the product. In this sense domestication is aimed at modifying the supply curve of the extractive resource according to the availability of technology, affecting the stability of the extractive economy, insofar as the success of the domestication of the product will cause the extractive sector to contract (Homma, 1989).

Alternatives to the model

One of the main problems in Amazonia is the large-scale devastation of the native forest cover. One of the main activities introduced after deforestation is extensive cattle raising.

It is estimated that with extensive cattle raising, 587 hectares are needed for each worker, which is 300 times more land than that used by a rubber tapper for small farming. In extreme cases, deforestation may be for the benefit of the timber trade, giving rise to social problems by expelling extractivists to urban peripheries.

According to a study by the researcher Suzanna Hecht, cattle-raising can only be extended for 10 years without additional costs. After the end of the tenth year, if the activity is continued, 270 dollars per hectare per year would need to be spent on fertilizers. Costs would increase considerably for the option of total reforestation (recuperation) to obtain native species, amounting to 7 400 dollars per hectare.

Thus there is a natural tendency to abandon the land completely at the end of the tenth year, considering that it is much easier and cheaper to clear away the forest on new land than to keep existing lands fertile. A proof of this is that 50% of the devastated areas have already been abandoned.

Several studies point to the non-viability of extensive cattle raising and itinerant agriculture, which generate extremely low (or even negative) returns per hectare and have been maintained by high subsidies (Buschbacher 1986, Hecht et al. 1988). In this connection, it is worth remembering that most of the soils in Amazonia are relatively infertile.

Considerations for the analysis of the economic viability of extractive reserves

— The microeconomic analyses made by critics of the extractive reserves forecast the interaction between the variables concerned, such as demand, price, supply, etc., without taking into account that exogenous events in this case are extremely relevant, more specifically the social, cultural and political factors, which are an integral part of the very concept of an extractive reserve.

— A common feature of all the criticism on economic grounds is the failure to incorporate the positive externalities generated by extractivism. Their identification, acceptance and appropriate remuneration constitutes one of the fundamental aspects of any policy aimed at promoting extractivism as a model for the conservation of natural resources.

— The evaluation of economic efficiency (measured in terms of dollars per hectare) produces distorted conclusions, since it does not take account of the value of biodiversity in the reserves, which represents a store of wealth which is not priced at market values, as opposed to the extracted products which are sold. As a result of this distortion, the use of 300 hectares of pasture employing only one person (and producing cattle) will be considered just as or more productive than the use of 300 hectares of land by a family of rubber tappers, producing rubber, but with 95% of the land occupied by forests with enormous biodiversity, in terms of both fauna and flora. This crucial deficiency of market mechanisms to indicate the best use of natural resources has already been amply recognized (Dasgupta and Heal, 1979; Pearce, 1976).

— In the case of forest areas, the available options include: large private land holdings (cattle farming companies), a multiplicity of small individual plots, also private ("colonization"), the system of reserves (public property combined with rights of use delegated to associations of rubber tappers), and national parks. Past experience has shown that the cattle farming companies and settlements by small landowners have potentially tragic effects on natural resources.

— Until now the social costs of migration to urban areas and the loss of traditional skills required for multiple uses of the forest are not sufficiently reflected in economic analyses.

— The low yield of extractive workers is also attributed to the type of marketing used. Extractive communities have been exploited in their trading relations (through the system of *"aviamento"* or debt peonage), with terms of exchange which were and still are in some regions unfavourable, in other words, the extractive workers receive extremely low prices for their forest products and pay excessively high prices for the market products they need for their subsistence.

— The local organization of extractive workers is improving the unfavourable terms of exchange. In the Vale de Acre, for instance, the *aviamento* system has disappeared, and nowadays many rubber tappers can market their forest products independently (Allegretti, 1989).

— When analysing the viability of extractive reserves, one should take account of the social changes generated by the creation of the reserves, which can improve and are improving the efficiency of extractive production.

— Critics who assert that extractive reserves are not sustainable owing to pressures on resources are disregarding the role of traditional institutions and new institutions in regulating access to resources. In fact, although ownership is public in the extractive reserves and use is guaranteed by government contract with associations, economic management and natural resource management are the responsibility of individual families, subject to rules preventing abuse (clearance control, hunting rules, ban on clearing areas containing rubber trees), within a larger settlement *(colocação)* inhabited by a group of families. Areas of common access and use are subject to the authority of the association.

— Strengthening the extractive reserves necessarily means strengthening the local community organizations, as the best way of implementing the joint management instruments legally constituted for the reserves. The best way of regulating the use of natural resources consists in strengthening self-regulatory mechanisms.

— It should be pointed out that the whole notion of extractive reserves is at present undergoing a process of enrichment and maturation, in which the actual extractive communities are also taking part. These communities, which initially were concerned only with land tenure problems, began to understand as time passed the need for a balance between environmental conservation, development and the adoption of appropriate technologies. To these new values were added others relating to the improvement in the quality of life, with their own ethical and moral values, related to the cultures of the communities involved.

— Critics who say that extractivism is technically backward often believe that forest dwellers specialize in the extraction of a single product. Quite on the contrary, the so-called "extractivists" display a remarkable variety of activities, which include commercial extraction involving sophisticated handling, but also hunting, fishing, agriculture, gathering and handicrafts (Almeida, 1992).

— The option of "multi-extractivism", or extractive activities diversified over many products and small quantities, provides practical solutions for concentration and domestication (Torres and Martine, 1991).

— The concept of extractive reserves should be understood within a broad dynamic context, which needs to be adapted to social and economic changes occurring in the world at large, making sure that the concept is never static in time or space. This obliges researchers and specialists to adapt their paradigms, if they are to understand the real potential of the scheme.

Proposed sustainable development for extractive reserves

The structure of extractive product marketing is undergoing changes, from one without circulation of money, in which rubber was the main product brought to market, to one in which the rubber tapper demands cash payment for his products, in order to acquire goods and services for which he previously had no demand because he was not aware of them.

Rubber is no longer the basis of the extractive economy, although it used to provide economic support for a large part of the population, owing to the fact that its price, underpinned by subsidies, guaranteed a steady income.

As an alternative, in order to ensure that sustainability is maintained in modern extractivism, the potential of the extractive economy needs to be increased, without detracting from its most basic quality, in other words, its ability to support and regenerate the resources it exploits.

In accordance with this objective, productive activities which are planned for extractive areas should be selected among those which are best adjusted to the particular stage of development of each one and to their communities' ability to assimilate and manage this increased potential.

In this preparatory phase of the Pilot Programme for the Protection of the Tropical Forests of Brazil, supported by the World Bank and by the Commission of the European Union, several non-governmental organizations, consultants and governmental institutions, including IBAMA, have carried out a number of studies aimed at establishing guidelines for a programme of sustainable development for the extractive reserves.

IBAMA/CNPT have supported the reserves' development in a number of ways; the results achieved have provided a range of experience, which in combination with the above-mentioned studies has served as a basis for general strategies in these areas.

In order to establish the strategies, some basic guidelines have been identified for any action to strengthen the extractive economy:

a) roposals should be made jointly, the bodies representing the reserve dwellers (associations, cooperatives and unions) being responsible for deciding, coordinating and supervising individual proposals and the process as a whole;

b) For all types of activity, objectives and implementing strategies should be made compatible with the the reserves' environmental management rules;

c) The development of activities must be technically supported by institutions working in the area, such as government bodies, independent researchers and locally active NGOs;

d) The training of the reserve population itself should be considered a basic element, including the training of local instructors to pass on technical know-how;

e) The development of productive activities should be open to other areas in the region, especially those in the vicinity of the reserves, in order to spread the benefits of internally developed results beyond their borders and avoid creating islands of development.

Some CNPT estimates indicate that the annual income of an extractive family in the state of Acre amounts to the equivalent of US$ 2 370 (including subsistence hunting, food crops, rubber, nuts, the rearing of domestic animals and the extraction of products such as *açaí, abacabá and patoá*), which represents a monthly income of US$ 197 per family. According to these figures, a rubber tapper's average revenue is more than double the "minimum wage" paid in the towns.

Nevertheless, if one considers that any purchase in the forest of goods imported from the towns (such as coffee, batteries, clothes, lamps, etc.) can cost up to five times their normal price, the situation appears extremely unfavourable.

The main points of the sustainable development proposal for the reserves are as follows:

1. Diversification of subsistence production
2. Fish breeding
3. Improvement of traditional products
4. Improvement of alternative products
5. Development of forest agriculture systems
6. Development of beekeeping
7. Rescue and development of traditional activities.

1. *Diversification of subsistence production*

It is very important to increase and to diversify subsistence production, since, in the absence of any consistent economic support activities, that production supplies the basic food content which can make family living in these areas viable and creates the right conditions for developing other productive activities.

a. Cleared land

In general, reserve families clear plots of land of up to 0.1 hectare, on which they grow their subsistence food, consisting of typical crops such as manioc and maize , in combination with a few other species, such as bananas, paw-paw, etc.

When a cleared area has been abandoned, generally only the pioneer species growing naturally are left in the soil. The extractive worker will return to the same area after a few years (rest period). In this cycle, there is no systematic form of management, so that the worker loses an opportunity to use a significant area as an extra source of subsistence and income. Moreover, very often areas are reoccupied before they have had time fully to recover (very short rest period), leading to lower productivity of crops and more labour costs per volume produced.

Management can take the form of enriching the cleared area (in the last year of planting) and other cultivated plots in general, introducing basic known and other food species to diversify and enrich the family diet, once the local cleared areas have been diagnosed and it has been decided what species to introduce jointly with the communities. Depending on how each family uses the particular area within the productive system of the settlement, other species apart from short cycle crops may be considered.

In cleared areas, some long cycle crops can be retained, alongside the newly introduced species. Paw-paw, banana, *inhame* and *cará,* for instance, will remain for a long period (up to six years) and can contribute to the family's subsistence.

For purposes of enriching cleared land *(roçados* and *capoeiras),* various legumes may be introduced, in view of their ability to fix nitrogen, which is important for general crop development. One species which is worth considering is the *ingá-de-metro,* which produces good quality timber and charcoal as by-products.

While adopting this system of management, an effort can be made to diversify crops in the cleared areas. Apart from basic plants such as manioc and maize, other species can be worked. Pineapple, sweet potato, *inhame, cará, taioba* (of the taro or *Colocasia* family), *jirimum,* watermelon, gherkin, red pepper, cucumber and tomato are examples of short cycle species, which serve the twin purposes of meeting short-term food requirements of the communities and enriching their diet.

Other food crops, which require processing before they can be consumed, such as sugar cane and coffee, can also be encouraged. These products come high on the list of the supplies for local communities, which are capable of achieving self-sufficiency in food within a short space of time.

b. Back yards or gardens (quintal)

In the extractive reserves, back gardens are generally small and the plants grown there little diversified, so that they do not produce much food value. Even local species are sometimes little cultivated or disseminated.

In back gardens, the rearing of small animals is more frequent than cultivation, although even it is little diversified and limited to the actual rearing activity (free range). Crops on cleared land are often grown far from the settlement centre, out of reach of domestic animals.

This free range breeding fails to take advantage of the manure produced, a rich natural input and a link between cultivation and animal rearing activities (for the recycling of nutrients), while it constitutes a threat to the family by transmitting diseases and worms.

The reason for free range breeding is supposed to be to save labour and because it makes it easier to feed the animals, since they themselves look for food in the forest, in the yard and in the near vicinity. The problem is that this system of breeding, particularly in the case of pig breeding, does not allow crops to be diversified. Enclosed, open air breeding, on the other hand, allows a variety of plants to be grown and makes the activity viable.

The breeding of poultry, another significant back yard activity, can also be diversified, with more fencing and chicken runs, built with posts and stakes taken from the forest. The most commonly bred poultry are chickens, followed by ducks; this activity can be diversified through indirect assistance, as there is no need to acquire breeding females and males to spread the activity among the families, which practically all have domestic animals. Other species which can be encouraged include turkeys, geese, wild ducks and guinea fowl.

In such units, the egg production could be sold by the families, with a view to diversifying their stocks. The eggs of any of these species can be hatched by chickens, which are generally good at rearing chicks which are not their own.

Diversification will have a direct effect on the forest fauna of the reserves, which is the main source of animal protein for most of the communities. Increasing the supply of proteins with domestic breeding will contribute directly towards restoring the stocks of the forest species most concerned.

As for diversifying back yard crops, the best is to develop suitable orchard and vegetable garden varieties, such as the *biribá* (Rollinia mucosa), mango (Mangifera indica), cashew nut (Anacardium occidentalis), *jambo* (Eugenia malaccensis), *acerola* (Malpihia punicifolia), oranges (Citrus aurantium), tangerines (Citrus nobilis), lemons (Citrus limonum), etc.

In enclosed vegetables gardens, the following species can be cultivated: greens *(couve)*, cabbage *(repolho)*, lettuce, onion (bulb and chibol), coriander, carrot, beetroot, turnip, red pepper, egg-plant, garlic, beans, okra, gherkins, cucumber and tomato.

The type of fertilizer used should always be organic, based on animal manure, the waste from plants grown on cleared land, such as rice and bean husks, and decomposing material in the ground cover (*serapilheira*). Once again, any production which is not consumed and not sold in these areas should be used as animal feed, thus completing an integrated production cycle.

2. *Fish breeding*

Fish breeding is an important activity which is worth developing in the reserves, for two reasons. Firstly, the areas involved offer favourable natural conditions for an activity which is little exploited, despite the fact that it is well suited to Amazonia and despite its economic and environmental potential. Secondly, expanding the activity could provide a significant extra source of income and food for the local communities, while contributing to a better management of natural fish stocks, which are at present under strong pressure from internal and external requirements.

Breeding fish in still water, on a scale sufficient to supply domestic needs, is a possibility which appears increasingly close to being realized in Amazonia. Since 1989, the Empresa Brasileira de Pesquisa Agropecuária (EMBRAPA) has been engaged in promising research work with three species of azolla, one of them native to the Amazon, an aquatic plant which grows in many tropical regions of the world.

Preliminary results of the research indicate that this plant can facilitate the reproduction and development of fish species which tolerate low levels of oxygen in the water. The plant also offers considerable potential as a producer of biomass with high protein content (up to 600 t/ha/year or 13.5 tonnes of protein/ha/year), a significant asset, in other words, as an organic fertilizer or animal feed.

These properties of azolla make it worth investing in the installation of tanks, for the reproduction and development, on a local scale, of the results already achieved through research.

In all the extractive reserves, moreover, there are natural lakes which can be managed to improve the conditions of fish production and even used to breed fish. Most of the lakes are directly linked to river beds, such as in the Alto Juruá reserve, where they are the result of the constant reshaping of the river Juruá all along its course, isolating meanders which then become horseshoe-shaped lakes. These lakes then provide areas of shelter and breeding grounds for aquatic species, including turtles, with good natural conditions for the development of aquatic fauna in general. If properly handled, these lakes can contribute a great deal to the production of fish in the reserves, with even the possibility of marketing.

In the Seringal Cachoeira, in the state of Acre, which was turned into an extractive settlement project in 1989, the breeding of *tambaqui* in small artificial lakes has helped to demonstrate the fish breeding potential of the area and to bring about a few changes in the eating habits of local communities, with favourable implications for the forest fauna stocks of the Seringal.

Managing the lakes can also be beneficial for the species living in them. In addition, the salting, smoking and manufacture of dried fish meal *(piracuí)*, amongst others, are ways of preserving the fish well known to extractive families and relatively easy to apply in the reserves.

3. *Improvement of traditional products*

The family income can be increased also by improving the processes of gathering and extraction of traditional products or introducing processing techniques which can add value to a product before it is marketed, either by the extractive worker or through cooperatives.

For the marketing of Brazil nuts *(castanha do Pará)*, a product which is readily marketable and is obtained by gathering the fruit of the Brazil nut tree (Bertholletia excelsa), the nut kernels are usually sold through intermediaries and are transported to places where wholesalers or factory buyers take them over. The nuts sometimes change hands several times before they reach a processing plant, and their price tends to vary according to the distance to processing centres. The losses in this traditional system of gathering, transporting and storing are relatively high.

Owing to the fact that the nuts are gathered from the ground, during the rainy season, the degree of contamination by fungi and other organisms is substantial. The gathering and transport of the product can take three or four months prior to processing. It is estimated that losses in the system currently run as high as 30%.

The best way to reduce these losses and to improve the quality of the product would be to reduce the time which elapses between the gathering and the processing. A system of decentralized processing of Brazil nuts aimed at reducing the losses and improving extractive incomes is at present being tested by the Cooperative Agroextrativista de Xapurí through pilot projects run in the communities (rubber and nut plantations) and in the actual places of residence of the extractive workers (allotments).

The cooperative, by improving processing and marketing channels, increased the market price of nuts paid to extractive workers by some 100%. This price increase benefits not only the members of the cooperative, but also other workers in the Xapurí region as a whole.

Other buyers, in competition with the cooperative, are forced to pay the same price or more. The price which the cooperative is able to pay is based on its selling price in the international market, less administrative and processing costs. It is thanks to the cost reductions achieved through decentralized processing that the cooperative has been able to double the price it pays to the nut gatherers.

In the rubber-growing areas, after collecting the latex the rubber tapper returns home, where on a different occasion he begins to smoke the "ball" of latex or to press the coagulated latex, producing CVP (Cernambi Virgem Prensado) or pressed virgin rubber.

In 1993, IBAMA started to distribute kits to produce PBD (Placa Bruta Defumada) or smoked raw sheets, with a view to promoting a better quality product with a higher commercial value. This technique, which is simple and can be used by the family, saves the rubber tapper's work time and avoids him having to be exposed to smoke when the latex "ball" is being processed, which has been a health hazard for these workers and has kept the quality low, with a direct effect on prices.

4. Alternative products

Rubber and nuts have traditionally been the reserves' main economic subsistence products. This meant that the extractive worker's income was always extremely dependent on a market which sometimes fluctuated considerably. In the case of rubber, the fall in prices which has occurred in the last few years has shown how fragile a system can be if based on a single product. There is a need, therefore, to introduce pilot experiments for the gathering, processing and marketing of new products.

The following objectives should be pursued: a) the identification and market testing of new reserve products; b) training reserve workers to handle the new resources, to manage them commercially and to acquire whatever know-how is necessary, and c) establishing criteria for and monitoring the sustainability of the new products.

Between 1975 and 1985, exports of extractive products increased, with a broader range of exported products and a rise in some of the prices of the more important productsl. It should be pointed out that, apart from native rubber in the post-war period, no product receives any incentives or subsidies and that efficient marketing networks are now operating in the international market.

Three products, nuts, palmito and carnauba, accounted for over 80% of the total value of exports in 1985. Carnauba (a type of wax extracted from the leaves of a palm tree used in the manufacture of cosmetics) is produced only in the north-east. Palmito planting was concentrated in the almost oligarchical *açaí* (Euterpe oleracea Mart.) forests of the Amazon river estuary, where the Cajarí reserve is situated.

Many of the species in the extractive reserves are used on a small scale for domestic consumption by the families and are not marketed, although there either is or there could be a regional, national or international market for such products. These include foodstuffs, oils, resins, fibres and essences (aromatic and medicinal).

Although a large number of species in the reserves are used, most of them are exploited only on a small scale, in ways which are not standard and with marketing channels which are precarious.

The most used species are the palm trees, other oleaginous species (*timbó*) and forest fruits. For instance, the *jaci, jarina* and *buriti* palms are used as roofing for houses; the trunk of the *paxiúba* provides a light, strong wood which is much used for house flooring and walls. *Açaí, bacaba, patauá* and *buriti* provide food. The seeds are left in tepid water and crushed to make "wine". If the "wine" is boiled, the oil can be separated off and eventually used by the extractive workers as a substitute for soy oil or as hair oil.

Palmito is obtained from the *murumuru* and from *açaí. Copaíba* oil is generally known as a remedy for inflammations, coughs, etc. *Timbó titica, timbó-açu* and *arumã* are used to manufacture sieves, baskets and brooms.

One species that deserves special attention is a type of laurel (*louro-de-óleo or pau d'óleo*). The oil extracted from this tree is also known as "gasoline", and is used to run petrol engines and in lighters and lamps. It is also taken as a remedy.

With these examples we can say that the challenge of diversification lies in taking advantage of the biodiversity of the forest, that is, turning the existence of many irregularly distributed species into an advantage instead of a commercial disadvantage, with a view to benefiting the inhabitants, without degrading the natural heritage.

5. *Development of forest agriculture systems*

An alternative to the development of reserves has been the establishment of "consortia" (*consórcios*) within particular areas, with a view to limiting as much as possible any disturbance in the families' working arrangements.

The areas which can be used for these consortia must be on cleared land (*"capoeiras"*); forested areas should be used only where cleared land is not available.

The consortia can serve as seed banks for the area, since many of the species are not cultivated in the reserves. The *pupunha* palm, for instance, is very suitable for the production of palmito, but growing it for this purpose requires a very large quantity of seeds, which the reserves are not yet able to offer. Thus the consortia can help to introduce species and support their dissemination using good quality genetic material.

For communities living on the upper river banks, the forest agriculture consortia should preferably be established along the rivers. In the areas where they operate, subsistence crops can be grown until the second year of operation, thanks to the good luminosity between the lines of cultivation.

As far as Amazon fruit trees with economic potential are concerned, the National Research Institute of Amazonia (INPA) launched a project a few years ago for the dissemination of crops of native species of the region, which are thought to be worth developing for their economic potential. Some 20 Amazon species have been worked on by the Institute, including: *cupuaçu* (Theobroma grandiflorum), *açaí* (Euterpe sp.), *bacabá* (Oenocarpus bacaba), *pupunha* (Bactris gasipaes), *graviola* (Anona reticulata), *camucamu* (Myrciaria dubia) and *sorva* (Couma macrocarpa), especially in the states of Rondônia, Acre and Amazonas; some of these were brought from other Amazon regions, while others were "discovered" among the Indians and rubber tappers, who use them as part of their diet.

These are species which, owing to their nutritive properties and taste, can eventually acquire market value, although that would take time. Even so, they could provide a useful addition to the food diet of local populations, and at some stage in the future could offer new options, improving the profitability of families and reducing the degree of risk incurred in their productive activities.

6. *Development of beekeeping*

Beekeeping is a type of production which is fairly well adjusted to the natural conditions of Amazonia and has been expanding in the region in recent years. As an activity, it is easy to manage and it has provided a good alternative in areas where the impact of the expansion of the "agricultural frontier" has been most felt, as in the case of Rondônia and other areas where colonization has predominated.

Moreover, with the forest agriculture systems and the improvement of cleared lands (*capoeiras*), introduced as a way of minimizing environmental impacts, it has been possible to add beekeeping as an activity and honey production has provided a very significant source of income for local people. Several specific beekeeping projects are being developed in the region, contributing to the diversification of productive activity and ensuring better conditions of sustainability.

A considerable variety of forest bees occur naturally in the region. The species most used for beekeeping in Amazonia are the africanized bees, which also occur commonly in the forest; this facilitates the establishment of hives in the reserves for a lower cost, through the capture of swarms in the forest achieved by the method of scattering bait-boxes in selected areas.

7. *Rescue and development of other traditional activities*

In all the reserves, as indeed throughout the rural environment of the Amazon, craftsmanship and home manufacture of objects in general are a common feature of family life.

Although they are important from a cultural and social development point of view, handicrafts, with rare exceptions, do not fulfill any significant economic function. Besides, with the recent rapid changes in consumption habits, combined with easier access to industrialized products, the traditions have gradually been losing ground.

On the other hand, hand-crafted products in the past have played (and to a certain extent still play) a part in local barter trade, facilitating a division of labour and production between the communities.

On the other hand, modern society as a whole has been conducive to a greater appreciation (even in aesthetic terms) of rustic and home-made articles, which give them a certain marketing potential.

Since these activities are traditional and carried out on a small scale, since they do not depend on outside investment or know-how and do not require training or maturation as a productive process, and since they do not represent any risk (economic or environmental) for reserve communities, any effort to rescue and enhance them should be given due regard in terms of adding to the potential of the extractive economy.

Strategies for success of the reserves

a. Increasing the output and productivity of existing forest products

Historically, the growth of "wealth" has been due more to an increase in productivity than to other factors. Productivity gains have outpaced the refutation of the "Malthusian

theory'' by the local population. Theoretically, there are no limits to productivity gains and increases of up to 100% have been the rule rather than the exception.
For extractive workers, the opportunities are the same as for other forest dwellers. An increase in productivity offers the best chance of higher incomes and a higher output, and productivity can be obtained by planting species to enrich the forest, by handling existing stocks properly and through technological improvements in harvesting.

b. Reducing forest product losses

The second most important factor for increasing economic gains is loss reduction and the elimination of inefficiency, which, however important they may be for the cost/benefit ratio, are not unlimited. The most that can be done is to eliminate 100% of losses, which will produce small gains, but will have a major positive impact on revenue and on marginal liquidity. The simplest and cheapest way for forest dwellers to increase their incomes is to reduce post-harvest losses.

c. Improving marketing in the local market

For the purpose of increasing the revenue of extractive workers, an efficient marketing system is essential. The system should be linked to an association or a cooperative. The two purposes of these organizations are to offer marketing services to their members and to deal with competition from traditional buyers.

The action of a cooperative or association in the region can become a critical factor in raising the prices paid for forest products and in reducing the prices paid for consumption goods by forest dwellers.

d. Adding value to the product and integrating production vertically through decentralization or local processing

Vertical integration and adding value to the product at the level of the extractive worker are other steps in the strategy of integration and income growth derived from forest products. Processing close to the place of harvest, in labour-intensive systems, is cheaper than processing in large factories, owing to the low opportunity cost of family labour compared to hired labour. There are other potential gains which contribute significantly to the family income, such as the reduction of losses and the lowering of transport costs.

e. Developing new markets for existing products

New markets need to be discovered for existing products. There are hundreds of products which are very well accepted in the regional market, but which are not known in other regions and much less in the rest of the world. This does not mean to say that all regional products will be liked, or that some of them will not be rejected by the market, but there is definitely a potential economic gain to be had from some of these products.

f. Developing markets for new products

There are a number of natural resources which are not being used but which need to be analysed. An assessment of their possible uses and of economic viability should be part of the strategy for saving and maintaining the tropical forest.

g. Making management operational

There is a need to strengthen compensation mechanisms in order to protect biodiversity. For this, a great deal of human and financial resources need to be invested in order to identify new products, to develop pilot experiments for product processing and regional, national and international marketing, to undertake applied research for continued identification of products, and to train the reserves' staff (in associations and cooperatives) in processing, quality control and marketing techniques. There is also a need to strength en the institutions involved in managing the reserves to improve the monitoring and evaluation of the ecological sustainability of new crops.

The purpose of this proposal is to increase the family earnings from the production and marketing of produce, to increase the productivity of subsistence farming through the use of selected seeds, and to introduce new techniques and suitable types of crops. Measures designed to improve handling, post-harvesting and primary processing conditions will contribute directly towards achieving the objective of increasing family income by making the extractive farmer more efficient in his agricultural activities, even when these are little complementary. Just as important as the increase in productivity are attempts to make marketing more efficient, an area where the participation of associations and cooperatives working in the reserves is absolutely essential.

The importance of technology

Technological improvements must be considered a fundamental part of extractive reserve development strategies. To this extent, there are three basic points which should be borne in mind with regard to the choice of technology: its socio-political significance; its scale, modernity and sophistication; and its probable environmental impact.

Vitor Bellia, a CNPT consultant, in his analysis of the technological issue proposes a set of criteria for a multidimensional analysis of technologies, which may be adapted to the conditions found in the reserves:

— economic efficiency
— operating scale
— degree of simplicity
— capital and labour intensiveness
— degree of environmental aggressivity
— finite resource requirement
— degree of autochthony and self-maintenance.

For a given technology to be considered appropriate, it should produce favourable effects (or the least unfavourable effects) with regard to the seven dimensions identified above. Clearly this is a matter of assessment, which will only be valid if it is part of an ideal concept of social systems. In other words, no technology can be appropriate in an absolute sense; on the contrary, it is bound to be either more or less appropriate to the extent that it enables the social system in which it is (or will be) used to come closer to or move away from its ideal characteristics.

a. Economic efficiency

To be acceptable, a technology has to be economically efficient, since it may be assumed that it will be applied in a diversified context, where the rules of the market place and competition play a central role.

b. Operating scale

The more the operating scale of a technology is compatible with the purposes for which it will be used, the more appropriate it will be. We reject the unconditional glorification of small scale, in view of the need for economic efficiency and the practical impossibility of resolving each and every problem by means of "small" solutions. We feel it is more realistic to advocate the adoption of technologies, the scales of which are compatible with their end-use.

c. Degree of simplicity

A technology has to be simple, that is, easy to understand and to use, without the need for esoteric knowledge or skills. There is no incompatibility between simplicity and sophistication. Simplicity is an attribute of appropriate technologies, while the lack of sophistication is no obstacle.

d. Capital and labour intensiveness

As a general rule, the most appropriate technologies will be those requiring a greater quantity of the least scarce factor, which means to say that when labour is readily available (which is practically always the case in developing countries), then labour-intensive technologies will be the most appropriate. Practice has shown that this observation, which appears so obvious, is frequently disregarded.

It is impossible, at the same time, to ignore the fact that products and processes need to keep a minimum standard of uniformity and performance if they are to be adopted and to become "socially acceptable". In many cases, the best solution is a combination of human labour and machines; in other words the choice is not so much between purely manual versus purely mechanical processes, but rather between different man/machine combinations. Subject to these reservations, it is fairly safe to say that, in developing countries, the more labour-intensive a technology is, the more acceptable it will be.

e. Level of environmental aggressivity

This is a criterion which does not need justifying, the rule being that the less a technology aggresses the natural environment, the more appropriate it will be. Non-violence is an irreplaceable and essential attribute of appropriate technologies.

f. Finite resource requirement

The more parsimonious a technology is in terms of finite material consumption , the more appropriate it will be.

g. Degree of autochthony and self-maintenance

It may generally be said that the more a technology depends on resources available in the actual social system where it is being or will be used, the more appropriate it will be. This rule should be applied with caution, owing to the problem of ensuring economic efficiency. Nevertheless, there is no doubt that it is generally preferable to use local solutions, independently of imports.

Autochthony is also an important factor for the preservation of local culture. It has to be borne in mind, however, that maintaining traditional cultural values should not become a pretext for social immobilism.

In production units, members share out their labour among different subsistence and extractive activities *(photograph: Julio Ruiz Murrieta).*

Marketing of the Brazil nut "Bertholletia excelsa" *(photograph: Julio Ruiz Murrieta).*

GUIDELINE FOR THE CREATION AND LEGALIZATION OF EXTRACTIVE RESERVES

Portaria IBAMA N° 51 de 11.05.94

This is to serve as a guide for the communities and the governmental and non-governmental organizations wishing to take part in the creation and legalization of extractive reserves. It sums up the experience acquired with existing reserves, and reflects the knowledge of reserve dwellers, who through their representatives have helped to prepare it. The guideline was approved by the IBAMA/CNPT Consultative Council at its fourth ordinary meeting on 25 February 1994. IBAMA made it official in its directive No. 51N of 11 May 1994, which was then published in the Di-87-rio Oficial da Uni-8b-o (Official Journal) on 13 May 1994.

1. Request by residents

The first step towards setting up an extractive reserve is for the local residents to make a formal application to create a Conservation Unit. Considering that the objectives of the extractive reserve can only be achieved if the efforts of the government and those of local residents are combined, it is essential that the latter, insofar as they will be in charge of the reserve in future, should take part in the establishment process right from the beginning.

The preconditions for ensuring satisfactory progress later on are firstly that the residents should agree among themselves and secondly that they should understand what they are requesting.

The application should be signed by the residents and by the bodies representing them or advising them (such as associations, cooperatives, unions, committees, etc.). If possible, the application should be accompanied by letters of support from other bodies (CNS, unions or non-governmental organizations) and by formal expressions of support by the local authorities (prefect, local councillors, etc.).The following information must be given with the application:

a) number of extractive workers residing in the area;
b) quantity of products of extractive origin (by crop or annual);
c) product marketing lines and prices;
d) proposed limits for the reserve (natural geographical boundaries, municipalities, districts);
e) residents' intentions regarding the use of natural resources.

The application is then submitted to the CNPT/IBAMA office in the relevant state, which will start a procedure for the creation of the reserve. If there is no state office, the application can go straight to the national CNPT/IBAMA coordination office.

2. Residents' associations

To set up a reserve, it is absolutely essential that the local residents should wish to work together, as an association. If there is not one already, the residents must set up an association which can represent them. They should all then work to strengthen the association and to achieve understanding, harmony and cooperation among themselves.

It is no good founding an association simply as a precondition to setting up a reserve. The association should be a natural outcome of the understanding or the perception on the part of the residents that they will have to be united and organized if they are to manage the reserve together. It is better in fact to work on strengthening the association before applying to set up the reserve, whose future is bound to depend on the association, to which the operating license (Concess-8b-o de Direito Real de Uso) will eventually be issued.

3. CNPT/IBAMA survey

Complementary area survey.

The CNPT state or national coordination representative will visit the area to check:

- the demographic situation and cultural traditions of the local population; this will include meetings with the residents to feel out the level of awareness of the problem, proposed solutions and the degree of organization;
- use of natural resources: residents' main occupation and sources of income; quantity of products extracted and prices obtained; future exploitation potential;
- non-extractive activities in the area, if any: farms, timber companies, trade, industry, etc.;
- the real estate situation in the area;
- existing infrastructure (transport, education, health, communications, housing, etc).

The survey should concentrate on checking that the conditions required for the creation of extractive reserves under Decree No. 98.897 of 30 January 1990 are properly met, as follows:

Article 1 Presence of an extractive community and possibility of self-sustaining exploitation, preserving renewable natural resources.

Article 2 Special ecological and social interest in the area proposed for the reserve.

The survey will produce a preliminary report, indicating whether to proceed any further.

4. Supplementary data

The preliminary report will be considered by a CNPT/BSB technical panel assisted by experts. If the panel is in favour of setting up the reserve, then further data will be requested by means of the following surveys:

I. Mapping and geographic data

a) Preparation of general area map;
b) Map of the area's location within the State of the Federation;
c) Map showing existing natural resources (if no other data are available, the RADAM Project map may be used). All possible maps, including hydrographic, vegetation, soils, etc.;
d) Descriptive survey of area perimeter.

II. Biological survey of area

a) Characteristics of dominant species;
b) Occurrence and incidence in the area;
c) Present and future productive capacity;
d) Data concerning the viability and establishment of sustained management of resources (economic and ecological);
e) Diversification of flora and fauna.

III. Socio-economic survey

a) Profile of existing community;
b) Number of families;
c) Cultural and economic conditions;
d) Inventories of existing improvements (by public or private enterprise);

f) Ways of handling the use of natural resources;
g) History of area occupation.

IV. Photographic survey
a) Evidence of works carried out by residents;
b) Physiographic aspects of area;
c) Anthropological aspects of area;
d) Other.

V. Legal formalities
a) Petition by the community requesting the establishment of the reserve;
b) Constitutional aspects in accordance with federal, state and municipal constitutions;
c) General considerations with legal opinion of the regulatory situation (state legislation);
d) SUPES/IBAMA certificates, recognizing the environmental characteristics of the area;
e) Copies of agreements and minutes of public hearings concerning the question of resources available in the areas and possible social conflicts;
f) Certificate by the State Secretariat for the Environment, where possible, concerning the dominant environmental characteristics (ecological zoning, reports, opinions, etc.);
g) If possible, a land tenure map by INCRA or the State Land Institute with a register of properties if applicable;
h) Copies of agreements, minutes of public hearings held, protocols of intention, etc.

VI. Final comments
All available technical and scientific bibliography should be collected, together with newspaper articles and scientific reviews concerning the social and environmental situation of resources in the area.

5. Approval by the commission and establishment of the reserve

Once the above formalities have been completed, the CNPT submits the file for the consideration of the Commission for Traditional Peoples (this is a commission set up within IBAMA to deal with matters specifically related to traditional peoples).

Once the commission's approval has been obtained, the CNPT prepares a draft decree setting up the reserve, with a statement of the reasons, and submits the documents to the office of the President of IBAMA, who then undertakes the necessary legal procedures with the office of the President of the Republic. The establishment of the reserve is by presidential decree.

6. Land registration and writ of entry

Once the decree establishing the reserve has been published in the Diário Oficial da União, IBAMA proceeds to legalize the land tenure, leading to a writ of entry in IBAMA's name.

Legalizing the land tenure consists in identifying all realties and persons included within the reserve, in order to evaluate the lands and improvements and compensate those who have to leave the reserve. Once all realties have been expropriated, the judge declares a writ of entry over the area in IBAMA's name. The land then becomes State property. Expropriation gives rise to the payment of compensation decided by the court. Such payments are made in cash.

7. Register of residents

Once a reserve has been created, IBAMA should begin to draw up a register together with area residents. This will actually consist more in updating the existing register for purposes of the future licence (Concessão Real de Uso), since an initial register, if only provisional, of residents should have been made at the time of the socio-economic survey.

8. Reserve Utilization Plan

Under the terms of the decree setting up a reserve, a licensing contract (Concessão Real de Uso), which must include a utilization plan approved by IBAMA, has to be issued before any operations can begin. The plan is an administrative means of checking that the use of natural resources is self-sustaining, will not harm the environment and will comply with existing environmental legislation.

In the light of land and socio-economic survey data, the reserve residents, on the basis of their knowledge and experience, must then submit a Utilization Plan, containing the following main points:

8.1 Aims of the plan

- This part consists in giving a brief description of the means to be used for maintaining the reserve as a unit designed for self-sustaining exploitation and the conservation of renewable natural resources, by the residents;
- It should be explicitly mentioned that the plan contains a listing of non-predatory activities adapted to the culture of traditional peoples and in harmony with Brazilian legislation;
- It is preferable to declare that the plan may be used as a reference to check that the exploitation of natural resources is self-sustaining, not harmful to the environment and in conformity with existing environmental legislation; and that the plan may be referred to by IBAMA to check observance of rules stipulated with respect to the environment.

8.2 Persons in charge of implementing the plan

This gives an outline of the existing or future organizational structure set up to implement and monitor the plan.

8.3 Human activities in the reserve

The plan should list human activities, mentioning the principles which will need to be respected, which activities can be deployed, which activities are not allowed, and always giving details of the quantities and forms of activity involved.

Although all the human activities take place in the "forest", taken in its broadest sense, in order to give a clearer idea of how to prepare the utilization plan, we may divide human activities into different categories, as follows:

1. Extractive and farming activities: these are understood to include all activities which the residents normally conduct;
2. New activities in the forest: these are new activities which are to be introduced, extracting new products in an ideal allotment, such as fruits, oils, essences and even timber in the ideal allotment attributed to each resident;
3. Activities involving fauna;
4. Activities in areas of common usage: it is very important to give a description of any activities planned for areas of common use.

8.3.1 Extractive and farming activities

Type of items to be described:

a) Type of extractive activity;
b) Ideal size for each family to exploit;
c) Form of exploitation (how it is to be implemented);
d) Responsibilities of each person in his area;
e) Precautions and care to be taken in the reserve;
f) Spatial division between extractive and farming activities;
g) Size of area used for farming activities;
h) Use of cleared and pasture land.

8.3.2 New activities in the forest

The Federal Constitution considers the Amazonian forest as a national heritage, and federal legislation describes forests as environmental assets of value not only to the region in which they are situated, but to the nation as a whole.

In view of the social and ecological significance they assume nowadays, independently of where they are situated, forests will have to fulfill a social function. The social function of the forest will be fulfilled when available natural resources are being properly used and when the environment is being preserved. In the specific case of the extractive reserves, this condition will be fulfilled in the event of self-sustainable use of renewable natural resources by the extractive communities.

According to the Brazilian Forest Code, forests have areas which are under permanent protection and which may not be cleared; these areas are known as "permanently protected forests" and they basically consist of natural forest areas around the sources and margins of waterways (article 2).

In permanently protected forests, wood and other forest products may be freely extracted. However, any destruction of or damage to the forest, or any usage which violates their declared purpose is considered illegal.

In the light of these basic rules, the utilization plan will need to regulate the methods of sustainable plant extraction, in order to ensure the renewal of depleted natural resources.

The plan will have to describe the extraction of other forest products, such as fruits, oil and essences, for marketing.

Should extractive reserve residents decide to use any other forest product, subsequently to IBAMA's approval of the utilization plan, they may only do so if they submit a simplified management plan, describing the sustainable production capacity of the forest, and in conformity with the utilization plan and the National Extractive Reserve Programme.

The simplified management plan has to be approved by the Association's General Assembly or, if appropriate, by the union.

In the specific case of timber extraction for marketing, the Utilization Plan will set out clearly what can be exploited. If permission is granted, then a specific management plan will have to be prepared for this purpose and approved either by the association, or by the union or by IBAMA.

Residents may extract wood for their own use, for the manufacture of tools or for constructions within the reserve.

8.3.3 Activities involving fauna

The Brazilian Constitution protects fauna against practices endangering their ecological function, such as: deforestation or destruction of their habitats, the total extinction of species and any practices subjecting animals to cruelty.

Fauna is understood to mean all the animals belonging to a particular region.

Forest fauna is considered by the federal legislation as a public asset; in other words, the animals have no owner and may not be appropriated, even if they are on private property.

Professional hunting is forbidden by the Fauna Protection Code. There is no rule explicitly forbidding subsistence hunting, but it is understood that existing restrictions apply whenever the ecological balance is endangered or the hunted animal is threatened with extinction.

The trade in animals of the forest fauna and of any products or articles implying their hunting, pursuit, destruction or capture is also forbidden.

Only the trade in wild animals from duly authorized breeding establishments is allowed. Also allowed, subject to IBAMA authorization, is the capture of eggs, larvae and young animals for commercial breeding, and the destruction of forest animals considered to be harmful to agriculture or to public health.

According to Federal Law No. 7.653/88, most of these offences against the fauna are considered crimes and are deemed unbailable.

Federal legislation prohibits the use of forest animals outside their natural habitat. It is considered an offence to use any forest animal, even if it has not been hunted or pursued.

The Fishing Code allows fishing for subsistence, commercial, sporting and scientific purposes. It is prohibited in periods when fish migrate to reproduce and, in still or territorial waters, during periods of egg laying and reproduction.

The law also prohibits the fishing of specimens which need preserving or are below a certain size, or in quantities greater than allowed, or by the use of explosives, toxic substances or illegal appliances, implements, techniques or methods.

The Utilization Plan must explicitly prohibit any type of hunting for commercial purposes, as well as the sale of hunting products inside and outside the reserve. It must also prohibit the entry into the reserve of professional hunters, or of any other persons who are not residents for the purpose of hunting.

The only kind of hunting allowed is for the protection of the cleared lands and domestic animals, or for subsistence, so long as it does not jeopardize the ecological balance and the species hunted is not in danger of extinction.

The use of dogs in hunting must be regulated. If allowed at all, it must be specified in what circumstances they may be used.

Fishing (or shellfish gathering) must be limited to the feeding requirements of reserve residents, and any professional fishing must be prohibited.

Any use made of natural substances, such as *timbó, assacú,* or *oaca,* or of any type of appliance in fishing must be clearly specified. The practice of any traditional method of the region which may jeopardize the ecological balance or threaten the extinction of fishes must be prohibited.

Regulations governing the breeding of domestic animals in the reserve will mainly affect cattle, horses, swine and sheep. The method of breeding (free range, pens or fencing), as well as the responsibility on the part of the owner of animals and which type of animals may be bred in the area should be specified.

8.3.4 Activities in shared areas

An area of common use is "a good not subject to individual appropriation on a permanent basis. These areas come under the combined notions of private property and possession for common use, in which a degree of solidarity and social cohesion should prevail, based on agreed rules guaranteeing the maintenance of the areas".

Thus common areas include rivers, lakes, pathways, beaches, banks and natural areas administered jointly by reserve residents, for collective benefit.

The customs and usage of the community must be respected when these areas are used.

If any doubt or conflict should arise in areas of common use, the Residents' Association or any other body set up by the residents shall investigate and resolve the dispute.

The inspection of common lands is the responsibility of the Residents' Association, which may delegate its powers to the Reserve Protection Commission.

The decision to open up any new pathways must be taken by the executive officials of the Association, subject to the approval of the General Assembly of the Reserve's Residents' Association, provided that it does not harm the ecological balance or does not violate individual rights.

8.4 Supervision of the reserve

According to the law, IBAMA is responsible for supervising the Conservation Units. However, even though IBAMA is the highest supervisory authority, since an extractive reserve entails joint management by the government and by society, every resident must act as a supervisor over his own area and over the reserve, being responsible not only for caring for his own land but also for checking that the rules of the Utilization Plan are being observed by all the residents.

The Union, the Residents' Association and other bodies operating in the area should also be responsible for supervising and monitoring the extractive reserve, within the terms defined in the Utilization Plan.

It is advisable to set up a "Reserve Protection Commission" and a set of internal rules determining its objectives, powers, membership, criteria for election or exclusion, duration of terms of office, etc. These internal rules may be annexed to the Utilization Plan.

Those responsible for managing the reserve may also set up other types of inspectors, such as forest inspectors, setting out their objectives, powers, number, appointment criteria, etc.

8.5 Penalties

Penalties include all types of punishment imposed on anyone in any way infringing the law, the contractual clauses, or the majority decision of a social group or body.

Penalties are applied whenever there is evidence of a failure to fulfil, either wholly or in part, what has been agreed for the implementation of the reserve's objectives, in the light of the deliberations of the General Assembly or the Association's management.

The different kinds of penalties include: warnings, fines, provisional suspension and termination of contract.

The punishment must be proportional to the offence, in order to prevent any abuses on the part of those empowered to apply the penalties, and subject to the democratic principle of equity.

The environmental policing of the reserve is the responsibility of IBAMA, since apart from the fact that the power of policing pertains exclusively to the public authority, it was laid down in the Decree of Establishment of the Reserves that IBAMA would be responsible for supervising the extractive areas and for ensuring fulfillment of the conditions stipulating in the licensing contract.

In the light of the above, the penalties referred to in the Utilization Plan apply only to reserve residents, while the authority in charge of applying them is specified in the Plan and in the Association's statute.

In general, however, if IBAMA should find irregularities in the reserve, it may apply penalties both to the Reserve Association and to residents, depending only on who is responsible for the fault.

The main points which should be referred to explicitly in the plan are:

a) The types of penalties (warning, fine, suspension and termination of contract);
b) The specification of wrongdoings and the related penalty (one type of penalty for each infringement);
c) Who is empowered to apply each type of penalty;
d) The body to which the person penalized may appeal if he believes that the sanction was unjust.

8.6 General provisions

This part should include any aspects which do not fit under the above headings.

We may mention a few examples of points which have to be covered:

a) Conditions governing any exchange of areas between residents;
b) The reserve's natural resources will be managed by the extractive reserve Residents' Association in conjunction with other bodies (CNS, unions, cooperatives, etc.);
c) A rule forbidding the commercial exploitation of mineral resources of the soil or subsoil of the reserve;
d) A requirement for a Preliminary Analysis of Environmental Impact before any major constructions are undertaken within the boundaries of the reserve, such as may bring about significant alterations in the environment;
e) The above item is to apply both in the case of works decided by reserve residents and in the case of constructions planned by public bodies;
f) The possibility of modifying the plan as requested by the Association, which has to submit the proposed modification to IBAMA for approval;
g) The possibility for the proposed alteration to be suggested by a group of residents (number to be defined), who would submit the proposal to the relevant General Assembly;
h) No proposed modifications to the Utilization Plan may run counter to the purposes of the reserve.

8.7 Approval of the plan by residents

Residents must approve the Reserve Utilization Plan, at a meeting of the body representing them. The minutes of the approval meeting must be annexed to the Plan.

9. Letter of assent to residents

In order to facilitate further development activities, IBAMA may supply duly registered reserve residents with a Letter of Assent, which is accepted by financial institutions.

The purpose of this document is to facilitate credit operations with banks, even before the operating licence has been issued, in the event that the latter is delayed for any reason. The letter of assent is issued subject to two conditions: a) prior approval of the Utilization

Plan fixing the limits and conditions of utilization, and b) completion of the residents' register to identify who is entitled to receive the letter.

10. Approval of Utilization Plan

The CNPT annexes the Reserve Utilization Plan to the file and forwards it to the Commission for Traditional Peoples for consideration. The commission then issues an opinion, which, if favourable, recommends approval of the plan in order to initiate procedures for obtaining the Real Right of Use Licensing Contract. The Utilization Plan will then be officially approved by the President of IBAMA.

11. Real Right of Use Licensing Contract

IBAMA starts formalities for the preparation of the real right of use licensing contract, to be signed with the Reserve Residents' Association. The latter then signs an "authorization to use" contract or any other legal document it deems appropriate with each resident.

According to Decree-Law No. 271 of 28 February 1967, (Article 7), the real right of use licence is defined as a "solemn contract, transferring as a real right the temporary usufruct, for a definite or indefinite period, over a public or private land, for specific purposes of urbanization, industrialization, building, cultivation or other utilization of social interest".

In other words, it is an administrative contract by which the public authority attributes an asset in its keeping to a private person, to exploit for a specific purpose. Since this is a contractual instrument, which gives details of the objectives of the extractive reserve, it provides a better guarantee and greater stability for the legal relationship established between the licensee and the public administration.

The real right of use is granted free of charge, though it contains termination clauses in the event of any damage to the environment or of a transfer of licence "inter vivos".

Use may be allowed for a definite or indefinite period, as established in the Decree establishing the extractive reserves (Decree No. 89.897 of 30 January 1990).

The time factor is of its essence, whether limited or indefinite. An indefinite period is understood to last about 10 or 20 years. The concession cannot be in perpetuity, since it would be annulled by the fact that its objective is a legal impossibility (Article 145, paragraph II, Brazilian Civil Code).

12. Extractive reserve development plan

Simultaneously with measures to improve the reserves, further studies should be undertaken to prepare a Development Plan, which should at the same time improve and refine the utilization plan.

These studies will concern the quantification of stocks of natural resources, production technologies, environmental impact, etc., and will serve to quantify the development plan and the effects of human activities in the reserve.

The development plan amounts in fact to a practical implementation of the utilization plan, incorporating any corrections or improvements which residents wish to add.

The development plan takes the utilization plan as its starting point, and in the light of early experience and discussions with the community, proposes changes and additional activities, saying how much, when, how, where and by whom they are to be carried out.

The development plan is prepared by the community, with the support of IBAMA and institutions advising the reserve; residents quantify their activities and undertake to perform them; it is therefore a document which is to be written and improved in the course of the first year of implementation of the project.

The development plan should include the following main headings:

12.1 Reserve management training

The work of the first few months after residents have officially taken possession (by writ of entry) should provide an idea of what sort of training is required to manage the reserve. A proper training plan should then be prepared, both for those in management positions and for residents in general. The plan should include details of the types of training required and how to implement them, the length of training, places, participants, instructors, support institutions and financial sources, as well as the teaching material to be used.

Training fulfils an essential function insofar as it enables the residents to be independent and to undertake their own development. Training must therefore be planned as a process which will permeate all the development stages of the reserve and which will include retraining and evaluation.

The training plan should cover all aspects of interest to residents, such as community organization and development, organization and management of associations and unions, productive activities, storage and marketing, new technologies, health, education, social and labour rights, the environment, etc.

12.2 Social and community organization

The development plan must give an account of what measures are to be taken and what activities initiated for improving the social and communal organization of residents.

It should contain a description of existing bodies, such as associations, unions, cooperatives, councils or committees, together with their functions and the relations between them.

The plan should also explain the tasks for which the body representing the residents will be responsible, the hierarchical order and rules to be observed.

It should indicate a mechanism for decentralizing social organization, so as to ensure participation by all, for instance, through units established in different areas of the reserve, with a representative authority, on the model of union delegations. The various units might take the form of cells, communities, communal canteens or something similar.

It is also important to establish how the community is represented outside the reserve, and who will be responsible for negotiating on behalf of residents. The plan must also spell out how important decisions affecting the community are to be taken.

The type of organization chosen should allocate responsibilities for the different areas of activities, as a means of avoiding an excessive concentration of tasks in few hands. For instance, each of the following areas should have a different person in charge: health, formal education, construction, maintenance of common areas, planning, informal education and information, production, marketing, management and accountancy, etc.

12.3 Reserve management

Planning

The development plan should indicate what planning system is used, ensuring community participation. Planning should take the form of annual work schedules. These should indicate the types of activities, the time taken to implement them, those responsible and assistants, the human, material and financial resources required, and the origin of the resources.

Support and supervision

The plan will give details of support and supervision facilities for:

- staff recruited by the associations;
- health and education instructors and teachers;
- organizations active in the reserve;
- constructions;
- assets acquired and other investments;
- planned group and community works;
- preparation and implementation of the Environmental Monitoring Plan.

Inspection

A system of inspection must be described, including participation by residents and the training of assistant inspectors. It is advisable to set up a Reserve Protection Commission, which will also be responsible for inspection.

Internal rules should be drawn up for the reserve, setting out the terms of observance of the utilization plan, establishing the system of inspection, the role of assistant inspectors, penalties and other related matters.

Assessment

The plan should also indicate how residents can assess the progress achieved in the reserve and how eventually to introduce changes of direction.

12.4 Production and marketing

The development plan should include production targets, areas to be planted, new cultures to be introduced, animals to be acquired and distributed, the purchase of seed, nursery plants, targets and tasks for fishing, beekeeping or any other productive activity.

All these activities have to be quantified and programmed in accordance with the region's agricultural calendar.

The chapter on production and marketing should describe the technical assistance received by the reserve from governmental and non-governmental institutions.

Another important aspect to be covered are the terms of reference for the studies to be undertaken with a view eventually to diversifying and increasing production.

Craft and cottage industry activities should also be quantified, as a means of increasing residents' income.

12.5 Housing, transport, health and education

Considering that the main objective of the reserve is to improve the living conditions of residents, the plan should set out appropriate activities and strategies in the areas of housing, transport, health and education. It should also indicate the technical and financial participation by specialized institutions, specifying the responsibilities of each one. It is very important that the government institutions responsible for each of these areas should be brought in right from the start.

12.6 Institutional support

The development plan should also refer to the type of cooperation to be expected from government bodies and all other institutions supporting the reserve. Support facilities should include technical assistance and rural development, training, advice and financial support. Responsibilities should be clearly set out.

The main objective of an extractive reserve is to improve the living conditions of residents *(photograph: Jeanine Levistre).*

The usage and custom of the community have to be respected in "common areas" *(photograph: Jeanine Levistre).*

MANAGEMENT OF EXTRACTIVE RESERVES

Carlos Aragón Castillo

> *"The aspects of the extractive reserve proposal which fascinate environmentalists the world over are the very ones which, in their own dimension and with the appropriate rationality, constitute the great "charm" of the proposal. It did not originate with technocrats and has not been dreamt up by "intellectuals" to benefit the "people", but instead creatively identifies the roles of the state and of civil society, while it stimulates the imagination of the most orthodox specialists, chiefly because, ever since the debate began on alternatives to tropical deforestation, no other proposal has come forward which so effectively expresses the set of parameters required to establish a balance between development and conservation in Amazonian policy."*
>
> Mary Allegretti

Background

World wide, ecological problems have attracted attention in recent years owing to the aggravation of environmental disturbances, which are becoming one of the current concerns of part of mankind, confirming the awareness that the future of human life will be conditioned by the ways in which nature and its resources are appropriated and used.

Throughout the history of relations between mankind and nature, the imbalance of the environment and the depletion of natural resources always appear as contemporary problems, brought about by the dynamic and the expansion of the development models adopted.

In the developed countries, environmental problems appear essentially scientific, insofar as the societies concerned go about defining their policies in the search for solutions, while disposing of the necessary material and financial resources to implement them. These countries concentrate their efforts on directed research aimed at discovering scientific and technical ways of adjusting their huge productive machinery to maintaining a limit balance with the environment.

In developing countries, on the other hand, the solution of environmental problems is intimately related to the eradication of poverty, hunger and unemployment, since these are the conditions which induce man to appropriate natural resources, with no regard for the perpetuation of the resource from which he is currently deriving his sustenance. It has now become imperative to try to reconcile respect for and conservation of nature with the rational socio-economic development dynamic, and to reject the notion that economic growth depends on an input of natural resources and environmental degradation. This means approaching the rational use of natural resources as part of a policy of sustained development. In order to achieve this objective, Brazil's environmental policy now insists on means of ensuring the definitive incorporation by society of the new approach to the use of natural resources as a basic condition for combining viable socio-economic development with conservation of the environment.

Brazilian legislation highlights the environment as one of the aspects of the social order, and attributes to the public authority responsibility for preserving and restoring essential ecological processes, for monitoring activities affecting environmental equilibrium and the quality of life, for preserving the diversity and the integrity of the nation's genetic heritage, for protecting the fauna and flora, for creating territorial reserves in need of special protection, and for promoting environmental education and awareness.

Through the Instituto Brasileiro do Meio Ambiente e dos Recursos Naturais Renov-87-veis (IBAMA) (Brazilian Institute for the Environment and Renewable Natural Resources), the government implements its environmental policy and activates the participation of governmental and non-governmental institutions on a national and regional level, with a view to transforming the state's attitude and society's expectations regarding the environment. This integration of environmental policy as part of overall and sectorial development policies in Brazil is also reflected in the revision of government legislation, which, in both public and private projects, subjects the use of forests to the responsibility of maintaining the integrity of rural areas, so as to guarantee the conservation of the ecosystem and local communities.

Legal provisions governing extractive reserves

The establishment of an extractive reserve takes the legal form of a decree signed by the President of the Republic. This procedure must originate in a request from the extractive group concerned, and the area affected must comply with three basic legal requirements, namely its extractive potential, the possibility of self-sustainable exploitation and the presence of a traditionally extractivist population (as described in more detail in the above guideline for the establishment and legalization of extractive reserves).

This procedure for the implementation of extractive reserves refers to a series of initiatives which have to be undertaken by the public authority, by the beneficiary population and by the institution set up to ensure the effective participation of the community throughout the procedure, that is, the body representing the residents (association).

Each of these actors is assigned specific duties and obligations in the establishment procedure, as follows:

IBAMA is responsible for:

a) legalizing land tenure;
b) preparing the Real Right of Use Licensing Contract;
c) preparing the Model Authorization of Use;
d) general inspection.

The beneficiary community has to:

a) constitute bodies representing the residents;
b) prepare the Utilization Plan and the Development Plan.

IBAMA, jointly with the representative bodies and with the CNS, will be responsible for:

a) the population register;
b) the socio-economic diagnosis.

The residents' associations will be responsible for:

a) signing the real right of use licensing contract;
b) issuing Titles of Authorization of Use to reserve residents.

Guidelines for the management of extractive reserves

Environmentally speaking, natural resources should be seen neither from the point of view of the exploiter, who wishes to appropriate their economic value, nor from the point of view of the conservationist, who wants them to be untouchable, but instead in the spirit of identifying possible changes and adopting whatever measures are required to ensure that the equilibrium is maintained and that renewable natural resources are preserved in harmony with economic development. This requires a new approach to the general rules governing the principle of use, to ensure that the benefits derived either from exploiting or from preserving the resources are distributed fairly, both to those who derive their subsistence from the resources and to urban dwellers who indirectly absorb the oxygen exhaled by the forests.

Moreover, because tropical biodiversity is so important, it requires a new approach to the public management of forest resources, to achieve a balance between patrimonial values and the benefits derived from using the resources. Management policies and models must therefore give careful consideration to the relation between man and the use of these resources.

From the point of view of sustainable development, the management of natural resources will necessarily require a thorough review of the methods and procedures traditionally applied in the planning and administration of such resources. The diagnosis and the current management of the environment must take account of the interests and rights of local populations.

Past experience has shown that the great majority of use and conservation measures, mainly in native tropical forests, undertaken without consulting local populations have run into serious problems and have frequently ended in failure. As an example, we might refer to the vast areas of forest management in Malaysia, the African parks, the urban green areas of Brazil and the recent occupation of Amazonia. On the other hand, bringing local people into the management of natural resources has proved to be not only a positive factor for management but also a strategy for a socially more equitable distribution of benefits, thereby contributing to sustained development.

Institutional framework for the management of reserves - the CNPT

According to the decree establishing the extractive reserves, the Instituto Brasileiro do Meio Ambiente e dos Recursos Naturais Renov-87-veis, known as IBAMA, is the government body responsible for establishing, developing and managing the reserves. In order to fulfil these obligations, IBAMA opted to set up an administrative unit specifically responsible for managing the reserves. This led in 1992 to the creation of the Centro Nacional de Desenvolvimento Sustentado das Populaç-9b-es Tradicionais, CNPT (National Centre for the Sustained Development of Traditional Communities).

According to a relatively flexible interpretation, traditional communities are considered to be those which, traditionally and culturally, base their subsistence on the extraction of renewable natural goods. The concept is left deliberately flexible owing to the considerable diversity of communities which can and must be attended to by the government. These range from the *babaçu* coconut breakers on the borders of Amazonia, to the fishermen and shell gatherers on the Brazilian coast, the *"kalunga"* herdsmen, made up of black fugitives from slavery, who have maintained their hidden *quilombos* in the interior of the state of Goi-87-s, and many others. A rigidly defined concept might have led to the elimination of some human groups, such as the *"kalungas"* referred to above, which would have been ethically undesirable.

Set against our own culture, these populations are immensely fragile. Our concepts of property, profitability and economic efficiency, linked to our supply of comforts, have already destroyed many of these cultures throughout the world, absorbing the remains as an outcast population in our cities, disregarding the accumulated knowledge which allowed them to survive for all those years, sometimes for centuries.

In the Rio de Janeiro Declaration , this form of knowledge was finally recognized in

Principle No. 22, which states: "Indigenous people and their communities, as well as other local communities, have a fundamental role to play in managing the environment and in development by virtue of their knowledge and traditional practices ..."

Further on, the same Principle establishes that "Governments should recognize and support, by appropriate means, the identity, culture and interests of these people and communities, and should empower them to participate effectively in promoting sustainable development".

The Rio Declaration, signed by more than 150 countries, gave the backing that was needed to support programmes for traditional communities. At the same time, it increased the responsibility of those involved in promoting such communities, whether working on behalf of government bodies or otherwise.

It is no easy task to harmonize the development of traditional communities, preserving their identity, culture and interests, in the face of the sheer diversity of individual characteristics of each community and in relation to our own society. There are no methodologies available as yet, and it may even be doubted whether there ever will be. Every group is unique and therefore "empowering them to participate effectively in promoting sustainable development", as the Declaration urges, requires an individualized approach, duly respectful of their individual customs and knowledge.

The traditional extractivist communities contributed towards and/or were mainly responsible for the preservation and conservation of a large part of Brazil's renewable natural resources, currently under IBAMA's direct responsibility. The recognition of this fact provides a new framework of reference for the management and administration of Direct Use Conservation Units, insofar as it added new elements to the strictly environmental parameters.

The creation of the CNPT, in February 1992, was preceded by intensive discussions regarding the adjustments which would be required to achieve the triple objective of development, cultural preservation and environmental conservation, in such a way as to avoid these objectives becoming mutually exclusive. The conclusion at the end of the discussion was that the free determination by the communities themselves of the development of their cultures set an ethical boundary to the CNPT's action. This boundary, apart from being ethical, is also scientific, in the sense that it provides a means of optimizing the preservation of the cultures and knowledge accumulated by the people concerned. This principle applies not only to the CNPT, but should also guide the action of any organization or individual foreign to the traditional communities, however laudable his or their intentions may be.

In December 1991, in the light of this notion of an ethical boundary, a further sustainable development concept, specifically targeted at traditional communities, was issued as follows:

> *"Sustainable development for traditional communities is considered to be a process of transformation whereby the exploitation of resources, the application of investments, the direction of technological development and institutional changes are harmonized, so as to strengthen the present and future potential of the environment, which supports the economic activities of these people, for the better fulfillment of their needs and aspirations, while respecting the free determination of the development of their cultural profiles."*

Clearly, promoting development in this way is no easy task and not at all repetitive, like an industrial process. It is not amenable to centralization, or to bureaucratization. It cannot depend on the action of governments or on their limited resources. The CNPT's purpose, as defined in its internal rules, accordingly considers a framework of possibilities, in the following terms:

> *"The purpose of the CNPT is to promote the preparation, introduction and implementation of plans, programmes, projects and actions requested by traditional communities, through their representative bodies, and/or indirectly through specially constituted government bodies, or through non-governmental organizations ..."*

The realization of the complexity of its purpose, combined with the recognition of the inefficient way organizations historically transmitted to beneficiaries the resources which were their due, led the CNPT to adopt a different structure within the Brazilian public administration. Its regulations, for instance, give priority to support and promotion, leaving the practical action to be taken by the beneficiaries themselves, who decide what, why, when, where, how and with whom action has to be taken. The CNPT remains responsible for assessing the requests submitted by local communities and for adjusting them to the resources available, for providing the resources directly to the bodies representing the beneficiary communities, for monitoring their application and finally for assessing the results.

The CNPT is also responsible for organizing meetings between the leaders of local communities and experts and consultants, as well as courses in such subjects as management and administration, with a view to preparing them to manage their own sustained development, as recommended in Principle No. 22 of the Rio Declaration. The objective is medium term, with the future aim of making these traditional communities developed and self-sufficient, completely independent of initially offered subsidies. The CNPT is lastly responsible for organizing, structuring and guaranteeing the legal and institutional framework required to fulfil the objectives of each of the beneficiary communities.

This special structure of the CNPT also affects its management. Even though it is directly subordinated to IBAMA's management, its decision-making structure provides for the institution of permanent councils, made up of representatives of the traditional communities, non-governmental organizations and IBAMA's own representatives; votes is single and the CNPT's chief is designated as a deputy of IBAMA's president.

CNPT structure

In functional terms, CNPT is a form of joint management between the representatives of traditional communities, non-governmental organizations and federal and state government bodies, organized as follows:

I - Collegiate bodies:
- a) Advisory Council
- b) Regional councils

I.I - Executive bodies:
- a) Central Office in Brasilia
- b) State representations

The Advisory Council is responsible for supervising activities and for proposing guidelines and policies for the CNPT's action, and for approving the appointment of the CNPT chief. It is made up of 15 members, including non-governmental institutions and representatives of extractive communities.

The regional councils monitor the CNPT's actions and activities at regional level. They are attended by IBAMA's representative in the region, by a representative of the state government and by representatives of other civil organizations involved in the Centre's regional plans, programmes and projects.

The executive offices have to implement the CNPT's plans, programmes, projects and actions. They consist of an "interdisciplinary group", which includes experts of the state governments, IBAMA experts, and experts appointed by the NGO Forum.

Despite the amount of work involved, the CNPT offices in Brasilia operate with compact teams. The staffing dimension was decided soon after the CNPT was set up, in order to avoid excessive bureaucracy in the relations with the reserve communities supported by IBAMA.

The importance of community organization

It is no good talking about the management of extractive reserves without first having a grasp of how community organizations work, and we cannot understand community organization without first recalling the history of the reserves themselves.

Until the end of the 1980s, rubber production was the main economic activity of extractivism and rubber was the most important forest product for extractive areas. For many years, its production and marketing were organized by the rubber bosses, and thanks to a favourable market, there was no need for any further mobilization or changes in the way things operated.

As time passed and as the crisis in the rubber plantations worsened, more and more functions were taken over by the rubber tappers. In some areas of the region (mainly Acre), whole rubber plantations became relatively independent and production and marketing activities were taken over by the rubber tappers, although in somewhat precarious conditions. Small cooperatives appeared in these areas and the population began to organize in a form of self-management in response to this greater degree of autonomy.

This set off a process of organization founded on experience and a tradition of working with rubber, a non-perishable product, easy to transport and with a guaranteed market, conditions which would have been favourable to the introduction of any type of productive activity in the forest.

When this process was already under way, the price of rubber collapsed, falling to a level where the economic viability of the great majority of extractive communities in Amazonia was no longer assured. Furthermore, market control mechanisms, which had also been in place for decades, were relaxed, with the result that the marketing of the product was no longer guaranteed.

Other factors came into play, such as land ownership and the conservation of natural resources, which led to the so-called *"empates"* or stalemate situations, in which the unions played an important role and which encouraged the creation of the present-day associations and organizations.

It is important to remember that, under Brazilian law, the associations play a fundamental role in the development of extractive reserves, considering that it is to them that the licence to use the areas is issued.

It is now becoming clear that these organizations need to change, as a result of a number of factors, such as recent social upheavals, the type of socio-environmental management of the reserves, the need to devise and develop management tools, the sustainable development programmes proposed for the areas and the tendency for the state's participation in the national economy to diminish, with the consequent rise of market forces.

The problem with these changes is that the extractivist cannot be guaranteed any period of grace to adjust to the new reality, which demands a more efficient and effective system of organization and administration, and, especially in the economic area, for him to seek alternative ways of improving and marketing his produce and to diversify his economic activity.

Without rubber as the main economic prop, the process of modernization will have to be based on other activities. In this "new" approach to modernization, community and management organization will have a decisive impact on the way production is organized in practice. In the first place, the range of activities to be encouraged as part of the policy of modernization must be as diversified as possible and as close as possible to the conditions currently prevailing for the self-management of the reserves.

In a kind of upward movement starting from below, the production volume and diversity required in these areas will only be achieved if the local communities manage to mobilize their own efforts to organize and market production. Like in a circular chain reaction, this requirement in turn gives rise to a need for internal supply strategies, in a situation which is already undergoing change, which again creates a need to achieve a better standard of both private and managerial organization.

On the other hand, the institutional improvement must be based on objective, professional preparation, achieved through cooperation. There is therefore a need in practice to develop a suitable methodology in the areas of strategic planning, administration and community organization, jointly with the local communities, their leadership and their representative bodies.

The conclusion reached in studies carried out by experts as part of the Pilot Programme for the Protection of Tropical Forests in Brazil was that some of the basic prerequisites for a plan to improve the management of extractive reserves now have to be given priority by local organizations and the CNPT, as follows:

a) The bodies responsible for managing each extractive reserve, on a longer-term horizon, must be the associations, made up of all reserve residents, to the exclusion of persons from outside the reserves. It is admitted, however, that existing representative bodies may be used, regardless of their territorial scope, provided that specific sub-units are set up to implement and operate the reserves;

b) Economic undertakings and representative bodies managing the reserves should aim to keep thier activites, that is, the actual management of the reserves and economic activities, separate, although in the initial stages of organization they may operate jointly. The economic institutions may, if convenient, include persons from outside the reserves;

c) The institutions representing the residents and the economic entities should be structured from the start to be self-sustainable.

Extractive reserve management instruments

The management instruments adopted for the management of extractive reserves are:

a) the Utilization Plan
b) the Development Plan.

These instruments were described in the guideline for establishing and legalizing extractive reserves.

Encouraging local people to participate in the management of natural resources has proved to be useful way of achieving a fairer distribution of benefits and thus contributing to sustained development (*photograph: Jeanine Levistre)*.

EXTRACTIVE RESERVE OF ALTO JURUÁ

Manoel Lima Feitosa

Geographic and environmental data

The Alto Juruá extractive reserve was set up by Decree No. 93.863 of 23 January 1990, with an area covering approximately 506 186 hectares. It is located at the western tip of Brazil and of the state of Acre, between 8 40' and 9 30' latitude south and 72 00 and 73 00 longitude west, in the municipality of Taumaturgo de Azevedo, situated at a distance of approximately 150 km as the crow flies from the municipality of Cruzeiro do Sul, which has a common border to the south with Peru. It shares the same hydrographic basin with several indigenous peoples, in the west with the Kampa tribe of the Amônea river, to the north with the Jaminawa-Arara tribe, to the south with the Kampa tribe of the Breu and Kaximinaua rivers and to the east with the Kaximinaua tribe, all of these in Brazilian territory.

The reserve may be accessed either by air, travelling from Cruzeiro do Sul to the municipality of Taumaturgo de Azevedo, or by boat, also starting from Cruzeiro do Sul, by the Juruá river, a trip which in an ordinary boat can take three to four days or only one day in any type of gliding craft.

The area of the extractive reserve of Alto Juruá is part of a biome, which from a macro-ecological point of view includes at least four types of dry land tropical forests, which are never flooded, one with dense vegetation cover and another open. The other two types of forest are linked to the hydrographic network: the periodically flooded alluvial tropical forest and the terraced tropical forest, apart from other less clearly defined formations.

The lowland forests occur on sedimentary material of the Solimões formation, eroded to produce a low relief with a predominance of hills and crystalline rock formations, rising to altitudes of between 200 and 500 metres. Dry open forests of palm trees and bamboo are to be found in the area.

The characteristic forest families are the Sapotaceae, Annonaceae, Leguminosae, Lecythidaceae, Euphorbiaceae and Tiliaceae. With more than 50% presence per hectare, the occurrence has been recorded of abiorana-seca (Pouteria laurifolia), *abiorana-vermelha* (Prieurella prieurii), *axixá* (Sterculia pruriens), *envira-preta* (Guateria poppigiana), *ingá-xixica* (Inga alba), *joão-mole* (Pisonia tomentosa), *mapatirana* (Pouroma sp), *matamatá* (Eschweilera matamata), *morácea-chocolate* (Pseudolmedia multinervis) and *pau dárco-amarelo* (Tabebuia serratifolia). The shrub stratum (sub-forest) varies in density according to the structure of the tree stratum, being denser in areas of bamboo and palm trees, and becoming less dense as the upper stratum closes over.

The dense tropical forest appears distributed in patches in the central and eastern region, surrounded by open tropical forest, this latter formation occurring more frequently. The alluvial type of tropical forest of the periodically flooded plain occurs in limited fashion along the rivers Juruá and Tejo in a band of varying width. The terraced alluvial open tropical forest is the one which occurs least frequently in the area of the reserve, being restricted to two patches at the mouth of the Tejo river (IDEAS, 1993).

The ecological rhythms of the region are very much marked by the rainfall, with an annual total exceeding 2 200 mm; the rainiest quarter is December, January and February, while the driest is June, July and August. As an annual average, relative air humidity is above 85%. Average temperatures also decrease between June and August and annual averages are relatively low for the Amazon, below 25 C for the year as a whole, with an absolute maximum of 38 C and an absolute minimum below 8 C. Average annual sunshine is slightly less than 1 800 hours.

According to the Koppen classification, the climatic type is AM. In the light of the Bagnouls and Gaussen classification, the region is mild subtermaxeric, with less than one biologically dry month per year (ISPN, 1994).

The area is exposed to the equatorial air mass throughout the year, to the Chaco low pressure tropical air, mainly in summer, and occasionally in winter to the polar mass. This air circulation produces a rainy tropical climate with short periods of mild temperature.

As a result of the climatic factor, in terms of high temperatures and high rainfalls, which occur practically all year in the region, the soils have developed under the action of intense intemperateness and leaching, giving rise to soils of unusual physical and chemical characteristics, obviously in association with the prevailing type of geological material and time (RADAMBRASIL, 1977).

The area of the Alto Juruá extractive reserve is situated within the red-yellow podzolic soil domain which covers approximately three-quarters of the area. The remainder of the pedological cover is made up chiefly of eutrophic *Cambissolos,* which are present mainly in the Tejo river basin. Red-yellow *Brunizens* occur in small proportion and were mapped as secondary units of association. In the flat valleys of the Juruá and Tejo rivers, some hydromorphic soils still occur, among which the eutrophic little humic gley soils with high activity clays and clayey texture is the most expressive unit in terms of area (IDEAS, 1993).

Historic aspects

The occupation of Acre, as of all the Amazon region, began with the search for forest products which could be sold abroad.

In the 19th century, latex began to be extracted by the Indians in the lower Juruá and in the Purus. This rubber was sold to merchants in Manaus and Belem who travelled on the rivers, also buying *erva-doce* (canary grass), cinnamon, turtle oil, *copaiba* oil, cocoa and sarsaparilla. Rubber was the main product.

With the drought in the north east in 1877 and 1880, thousands of north-eastern rural workers, suffering from hunger and out of work, left looking for a better life. Many decided to try their luck in the rubber plantations of Amazonia.

After the bankruptcy of the rubber barons and the import-export companies of Manaus and Belem, the former employees and managers began to operate an itinerant trade as local bosses, some of them taking up residence in the area where they used to work previously. Food imports were replaced by local production of *farinha* meal and other products. The rubber tappers began to have larger families, taking up hunting, fishing and the gathering of other forest products. These were sold either through local bosses or through the *regatões* (itinerant traders).

At the time of the Second World War, synthetic rubber made its appearance on the market, partially replacing natural rubber. In Amazonia, between 1943 and 1985, the federal government began directly to support and stimulate the rubber production of native rubber plantations and price controls were introduced.

At the same time, the Instituto de Colonização e Reforma Agrária validated land claims in Acre. Federal banks financed the rubber growers with subsidized interest rates. This policy did not benefit the rubber tappers, and instead generated a reaction by the industrial sector against all form of extractive activity.

In 1985, when monthly inflation rates reached double digits, the government abandoned its policy of supporting the real prices of natural rubber. Long-term advances were suspended. The rubber growers of the region began to take an interest in timber extraction. In 1986, a debt collection expedition was set up with police support, causing a revolt among rubber tappers and residents in the Tejo river and Restauração rubber plantation area; after a demonstration by the local people, the government withdrew the police force. That same year and the following year, the rubber tappers were very apprehensive on account of research into timber potential and preliminary contacts for installing the necessary infrastructure. This was the immediate background to the project for setting up an extractive reserve in the area, an idea which was ventilated for the first time in 1987, in trade union meetings of the Restauração rubber plantation, at a time when the National Council of Rubber Tappers had organized a meeting of rubber tappers in the local area as part of its schedule of activities.

In 1988, when the National Council of Rubber Tappers began its activities in the local area of the Alto Juruá, a preliminary survey was made in the Tejo river basin and a project was launched to set up the first extractive reserve in Brazil. In 1989, the Procuradoria Geral asked the Empresa Brasileira de Pesquisa Agropecuária (EMBRAPA) (Brazilian Farm Research Enterprise) for an ecological qualification of the region and, on 23 January 1990, the Federal Government, in its Decree No. 98.863, established the extractive reserve of Alto Juruá.

Socio-economic data concerning the reserve

The extractive reserve of Alto Juruá, according to a cadastral survey carried out under the terms of the agreement signed between the Brazilian Institute for the Environment and Renewable Natural Resources (IBAMA), the Association of Rubber Tappers and Farmers of the Extractive Reserve of Alto Juruá (ASAREAJ) and the National Council of Rubber Tappers for the region of the Juruá valley, has 865 houses with a population of 5 821 persons, which gives an average of 6.73 persons per family, in 325 settlements (*colocações*) and 10 farms.

Most of the residents were born on the rubber plantations, and only a few, mainly on the banks of the Juruá river, were born in towns. Most are from the north east or descendants of north-eastern migrants, who arrived in the region during the first rubber cycle (1885-1912) or during World War II (1939-1945).

Thc area which is now the extractive reserve was inhabited until the last decade of the 19th century by indigenous peoples of the Pano language family, represented by the Kaximinawa, Jaminawa, Amoaca and other tribes.

Population density is 1.15 inhabitants per km^2. The reserve area covers most of the present municipality of Taumaturgo de Azevedo, in which at least 60% of the population resides.

The population is made up of 54.6% males and 45.4% females; about 56% are under 15 years of age; the school-age population, in the 5 to 15 year-old range, accounts for 37%, with 2 265 children.

Women leave home earlier because they tend to marry between 12 and 16 years of age. Men leave home later, tending to work for their parents until they are about 20 years old.

1991 socio-economic survey data indicate that 53% of the reserve territory is appropriated as an extraction area, while 47% serves either as a buffer between settlements or consist of uninhabited areas between rivers. Places with permanent inhabitants are identified by small clearings opened in the forest for the main housing or for small cleared fields or pasture lands. The reserve territory is divided as follows: 1) areas with strong human activity; 2) degraded areas with pioneer vegetation; 3) forest areas with extractive use and regular hunting; and 4) areas with very low and irregular human activity.

Migration within the reserve is observed to be very low. Families have been living in the same rubber growing areas for more than 20 years and the same settlements for more than 10 years. Moves, when they occur, may be due to a number of causes, such as marriage, pressure on natural resources or the increasing number of houses within a settlement, perhaps the wish to live closer to relatives, or the need for schooling or medical support, or even for better transport facilities.

All transport is by boat or by canoe. There are no roads, not even for pack animals. Communication within the area is by trails which are maintained by the residents themselves. The journey from the most distant points of the reserve to the headquarters of the association might take up to two days by canoe, not counting the stretches which have to be covered on foot.

In the educational sector, the 1991 socio-economic survey indicates that illiteracy is high, affecting 78% of the population, including 96% of children between the ages of 7 and 10, 78% of children between the ages of 10 and 14 and 71% of young people between the ages of 14 and 19. Illiteracy is greater among men than among women.

Again according to the 1991 socio-economic survey, 19 schools were identified in the reserve area and it was found that not more than one third of the child population of 2 265 children, in the 5 to 15 year-old age bracket, had physical access to schools within a radius of up to two hours walking from their homes. The schools are public, with teachers paid by the secretariat of education of the state of Acre, or by the education secretariat of the municipality. Some of the schools cannot open for lack of teachers.

In the area of health, the population lives without any sanitary infrastructure. Public health services are incomplete and infrequent. According to the record, between 1981 and 1990, the SUCAM visited the Tejo river area only once, for malaria prevention, although such visits are supposed to be made each year. Until 1991, the rubber tappers used to buy medicines from bosses and "*marreteiros*" (itinerant traders), or they would consult quacks or take home remedies.

In 1991, the Association of Rubber Tappers and Farmers of the Extractive Reserve of Alto Juruá signed an agreement with Health Unlimited to secure assistance in the health area. Twelve stations were built to receive residents; 12 health agents were trained and are supervised; 2 manuals were produced and a programme of regular vaccination was initiated. Facilities for the operation include an equipped boat and a permanent team including a doctor and a dentist.

The diagnosis prepared by Health Unlimited, according to the socio-economic report, included the following comments:

"Yellow fever is not a serious problem as in other eastern areas, but the vivax and falciparum malaria varieties are common. Hepatitis is a serious problem; the lethal delta variety was recently detected in the region. Besides this, the Juruá valley has one of the highest rates of leprosy in the world (almost 20 cases per 1 000 inhabitants). Neo-natal tetanus is one of the main causes of child mortality, and measles and flu epidemics cause frequent deaths among young people. (...) The highest rate of morbidity is due to problems of an environmental origin, such as infestation by intestinal parasites or skin diseases. Complaints of diarrhoea resulting from contamination by water sources are extremely common, and the imminence of contamination by the cholera epidemic originating in Peru and present in the north of Brazilian Amazonia is a constant source of concern. (...)" (Health Unlimited, 1992).

Management and social organization of the reserve

The area is managed by the Associação dos Seringueiros e Agricultores da Reserva Extrativista do Alto Juruá, the rubber tappers and farmers association, legally established with a statute and a properly elected executive body, which was founded in 1988 and currently includes 371 members. In 1992, as part of its managerial duties, its general meeting set up river committees to achieve better organization on a local level; in some areas, significant participation of residents was achieved, such as in the Rio Tejo, Rio Bagé and Boca do Rio Tejo areas; elsewhere, such as in the affluents of the Juruá river or on the river itself, participation has been low or non-existent.

The association's headquarters at Boca do Tejo boasts an infrastructure made up of a 300 m2 warehouse, a lodging area of 450 m2, a shop, a meeting room and office, an electricity generator, and a solar panel and battery to supply the radio transmitter. It also owns three whaler-type boats of 3 to 6 tons equipped with 18 hp diesel engines. In the interior of the reserve, there is a network of 7 stores measuring between 40 and 70 m2, equipped with scales and calculators. There is a fleet of 11 canoes, each with a load capacity of 500 kilos, equipped with 5 hp diesel engines, to cover the interior areas of the reserve. The association also has a set of rollers and metallic presses to produce smoked or clear sheet rubber, as well as a store in Cruzeiro do Sul, which was provided by the Brazilian Institute for the Environment and Renewable Natural Resources (IBAMA) and which is currently used by a number of associations. Three radio receiver-transmitters are installed and fully operational, one at Boca do Tejo, one at Restauração and one at Breu, run on solar batteries, which are in daily touch with the headquarters of the Juruá valley regional branch of the National Council of Rubber Tappers, at Cruzeiro do Sul. These radios provide essential support for commercial activities as well as for health and education services.

The activities of the families living in the reserve area consist in the extraction of latex, agriculture, hunting and fishing, handicrafts and services. About 65% of the families exploit the rubber trees and more than 90% have cleared plots where they grow manioc, maize and other products.

Work in the family tends to be divided along gender lines, with the men looking after the extraction of rubber, hunting and fishing and women working at home and in the vegetable garden. Both take part in agricultural work.

With regard to property rights, both women and children may own goods within the home or the rubber plantation. Rights in the division of investments are considered.

A rubber tapper normally works 267 days a year, spending 150 of those days on the rubber trails and 117 clearing the land and preparing meal flour. Families with 1.2 adults working in this way can produce between 1 200 and 1 800 kilos of *farinha and* between 960 and 1 200 kilos of rubber per year. The revenue obtained by marketing these products is used by the head of the family to purchase basic products such as salt, soap, ammunition for hunting, engine fuel for transport and for preparing the meal, as well as work, hunting and fishing tools such as axes, machetes, knives, blades to cut the rubber trees, shotguns, engines for making meal, nylon fishing nets, and articles of personal consumption, such as cloth, hammocks and other sundries.

Utilization Plan

It should be emphasized that once an area has been expropriated, it becomes State property and the usufruct thereof is granted to local extractive residents through their associations, which is a way of avoiding a fragmentation of the land. The precondition for making the concession effective is approval of the utilization plan.

At present, the Alto Juruá extractive reserve has a proposal, by the community, for the use of the area.

The association, which is responsible for administering the reserve's natural resources, has the right to supervise the conservation of the rubber trails, the stock of palm trees and other resources in accordance with the "Utilization Plan" previously approved in a general meeting. The association is also responsible, with the support of the National Council of Rubber Tappers, for planning and implementing development strategies favouring conservation and a strengthening of the association itself, with the help of equipment, infrastructure and the recruitment of technical staff.

Current projects

A number of projects are currently underway to support the extractive reserve of Alto Juruá. The main one of these is to include it in the Pilot Programme for the Protection of Brazilian Tropical Forests (Extractive Reserves Project), the general objective of which is to test appropriate models of economic, social and environmental management, with a view to improving the methods and procedures used by traditional residents, by:

i) strengthening the organization of production and of the community;
ii) developing technologies to produce subsistence goods and products for market;
iii) developing the economic and social infrastructure;
iv) maintaining stocks of natural resources;
v) preserving biodiversity.

Current projects also include:

Support for community organization

Organizing the extractive community was always a challenge, since the isolation of people in the forest and the lack of communication with urban centres are basic factors of life for extractive workers in the Amazon.

Communities began to have a say in government policy-making in the 1970s and this participation has increased ever since. In the specific case of extractive reserves, their viability must depend on sound organization, since a condition of the establishment of these areas is precisely joint management between the government and an organized civil body.

Training to improve rubber quality

In order to increase the rubber tappers' income, different kinds of training are being offered to improve rubber quality. The aim of the training is to introduce technology for producing smoked raw sheet rubber instead of the traditional system of producing rubber in balls.

Training for assistant supervisors

Because the extractive reserve is a direct use conservation area, which means that the community is involved in its management, training is now being offered on technical, legal and administrative aspects of supervising the area. This procedure is in accordance with Brazilian environmental legislation, under the terms of the National Council for the Environment's Resolution No. 003, of 16 March 1988.

Support for the improved use of cleared land

In forest areas affected by man's activities, a few measures are being taken, one of which is the establishment of an agreement between the Brazilian Institute for the Environment and Renewable Natural Resources (IBAMA) and the Association of Rubber Tappers and Farmers of the Extractive Reserve of Alto Juruá (ASAREAJ), for the introduction of model fruit and vegetable gardens to improve the use of cleared lands.

Support for Research Project

A technical and scientific cooperation agreement was recently signed between IBAMA and the University of Sao Paulo (USP), for the purpose of developing studies, research, the exchange of information and mutual advice for the implementation of activities in support of the Alto Juruá extractive reserve. Research work is currently receiving financial support from the John D. and Catherine T. MacArthur Foundation.

Main problems

The reserve's main problems at present are connected with health and education.

In the health area, although there is an official agreement, which is financed until 1997, with Health Unlimited to develop a series of projects, there has been a lack of either preventive or remedial action on the part of the public health services against the most common illnesses in the area, such as yellow fever, malaria, hepatitis, worms and leprosy.

In the education sector, problems begin at the school planning stage, since there is no school programme which is adapted to the seasonal nature of local productive activities, and especially not to the peak periods of planting (August and September) and the fortnightly preparation of meal. Apart from this problem, a lot of basic equipment is lacking, such as desks, slates, water filters, toilets, and teaching material such as books, textbooks and pencils.

The teachers are all women, young and with only basic training. There is no inspection within the reserve area. There is a shortage of schools and teachers in places in the interior of the reserve to cater for some 1 500 children who live outside the attendance radius of the main schools.

In the primary sector, there is no real credit and incentive policy to improve and increase the production of natural rubber from native trees, or any incentive to develop other extractive activities.

Additional socio-economic data

Production, prices and average gross annual income

Alto Juruá Reserve	Production	Price	Gross income	
	KG	US$	US$	Participation %
Manioc	2,761,080	0.03	82,832.40	7.78
Maize	2,422,000	0.12	290,640.00	27.29
Tobacco	27,853	0.38	10,584.14	0.99
Rubber	365,722	0.60	219,433.20	20.60
Poultry	6,660	2.00	13,321.00	1.25
Pigs	2,422	14.00*	33,908.00	3.18
Cattle	2,422	28.00*	67,816.00	6.37
Hunting and fishing	753,242	0.46	341,491.32	32.53
TOTAL			1,065,026.06	100

* Price per *arroba* (approx. 15 kilos)

Source: World Bank - IBAMA/CNPT - ECOTEC: Estudo de Pré-investimento de Projeto RESEX, 1994.

Production, prices and average gross annual family income

Alto Juruá Reserve	Production	Price	Gross income	
	KG	US$	US$	Participation %
Manioc	4,560	0.03	136.80	7.78
Maize	4,000	0.12	480.00	27.29
Tobacco	46	0.38	17.40	0.99
Rubber	604	0.60	362.40	20.60
Poultry	11	2.00	22.00	1.25
Pigs	4	14.00*	56.00	3.18
Cattle	4	28.00*	112.00	6.37
Hunting and fishing	1,244	0.46	572.24	32.53
TOTAL			1,758.92	100

* Price pe *arroba* (approx. 15 kilos)

Source: World Bank - IBAMA/CNPT - ECOTEC: Estudo de Pré-investimento de Projeto RESEX, 1994.

Residences and population per area - Alto Juruá Reserve						
Establishments	**Margin**	**%**	**Interior**	**%**	**Total**	**%**
Households	222	26	632	74	854	10
Localities	37	12	278	88	315	10
Farms	9	90	1	10	10	10
Population	1,482	28	2,832	72	5,314	10

Soure: World Bank - IBAMA/CNPT - Almeida, Mauro/Levantamento Sócio-econômico Projeto RESEX, 1994.

Sex of heads of household - Alto Juruá Reserve		
Sex	**Number**	**%**
Male	785	94.9
Female	42	5.1
Not available	30	—
TOTAL	**857**	**100.0**

Source: World Bank - IBAMA/CNPT - Almeida, Mauro/ Levantamento Sócio-econômico Projeto RESEX, 1994.

Age of heads of household - Alto Juruá Reserve (1991)			
Age groups	**Number**	**%**	**Cumulative %**
Below 30	246	31.1	31.1
30-40	241	30.5	61.6
41-50	135	17.1	78.6
51-65	122	15.4	94.1
Over 65	47	5.9	100.0
Not available	66		
TOTAL	857	100.0	

Source: World Bank - IBAMA/CNPT Almeida, Mauro/Levantamento Sócio-econômico Projeto RESEX - 1994

Civil status of heads of household - Alto Juruá Reserve (1991)			
Civil status	**Number**	**%**	**Cumulative %**
Religious marriage	514	65.6	65.6
Civil marriage	10	1.3	66.9
Cohabitation	177	22.6	89.5
Civil and Religious	11	1.4	90.9
Widower/separated	56	7.1	98.0
Single	15	1.9	100.0
Not available	74	8.6	
TOTAL	857	100.0	

Source: World Bank - IBAMA/CNPT - Almeida, Mauro/Levantamento Sócio-econômico Projeto RESEX, 1994.

Household distribution by number of children - Alto Juruá Reserve		
Number of children	**Number of households**	**%**
Not available	107	—
01 child	60	8.0
02 children	68	9.1
03 children	64	8.5
04 children	64	8.5
05 children	74	9.9
06 children	76	10.1
07 children	78	10.4
08 children	61	8.1
09 children	67	8.9
10 children	42	5.6
11 children	41	5.5
12 children	17	2.3
13 children	20	2.7
14 children	5	0.7
15 children	4	0.5
16 children	4	0.5
17 children	3	0.4
19 children	1	0.1
22 children	1	0.1
TOTAL	857	100.0

Source: World Bank - IBAMA/CNPT - Almeida, Mauro/ Levantamento Sócio-econômico Projeto RESEX, 1994.

Illiteracy rates by region and age group - Alto Juruá Reserve (1991)					
%	Tejo/Margin	Tejo/Centre	Juruá/Margin	Juruá/Centre	Reserve
Aged 5-9	94.4	98.5	93.3	98.6	97.0
Aged 10-14	87.6	88.9	76.5	87.6	84.5
Over 14	78.2	81.2	62.8	75.5	74.8
Total (+5)	83.1	85.6	72.1	81.8	80.7

Source: Almeida. Mauro/Levantamento Sócio-econômico da Reserva alto Juruá, 1991.

Illiteracy for the country as a whole (total and rural) (in millions) (1985)			
Regions	Literate	Illiterate	Total
Brazil (millions)	69.7	32.7	102.4
%	68%	32%	100
Rural (millions)	15.3	17.0	32.3
%	47.4%	52.6%	100

Source: IBGE, Anuário Estatístico do Brasil - 1986, Cap. 12,2 (p. 162).

School attendance for population between 7 and 14 years of age - Alto Juruá Reserve					
	(1) 7-14	%	(2) Total	%	(1) / (2) (%)
School attendance	392	31,2	497	9.3	78.9
Non attendance	864	68.8	2,734	84.6	31.6
Subtotal	1,256	100	3,231*	100	
Total	1,256		5,317		23.6

Note: Persons aged 7 or more, including no replies.
Source: Almeida Mauro, Levantamento sócio-econômico da reserva do Alto Juruá, Pesquisa, 1991.

Until the last decade of the 19th century, the area which is now the extractive reserve of Alto Juruá was inhabited by Pano-speaking indigenous communities *(photograph: Julio Ruiz Murrieta).*

CHICO MENDES EXTRACTIVE RESERVE

Manoel Lima Feitosa

Geographic and environmental data

The Chico Mendes extractive reserve was established by Decree No. 99.144 of 12 March 1990 and lies between the parallels 10° 05' 41'S and the meridians 67°- 56' 10'WGR and 69°- 48' 22'WGR. It is located in the state of Acre between the municipalities of Assis Brasil, Brasiléia, Xapuri, Sena Madureira and Rio Branco, according to the former political division of the state, within an area of 970 570 hectares. The area was mapped out by the army, through the Automated Mapping Centre (CCAUEX), under a technical cooperation agreement with the Brazilian Institute for the Environment and Renewable Natural Resources - IBAMA.

The Chico Mendes reserve may be accessed by road to the south-western tip, at Assis Brasil, along highway BR-317, except in the rainy period, when access is restricted only to the south-eastern tip, at Brasiléia. This BR-317 highway practically runs round the area of the reserve on its east-south side, with regular traffic all months of the year, but only as far as Brasiléia.

By river, it is possible to enter the reserve area by the Xapuri river and its affluents, except in the dry period, when there is practically no navigation owing to the low level of water and the emergence of very large sand banks. Another possible access is by the more eastern part of the reserve, through the municipality of Sena Madureira, by the Iaco and Macau-8b-rivers and their affluents. By air, all the towns around the reserve are reachable, provided that they have landing fields. The best access is through the international airport of Rio Branco.

The Chico Mendes reserve area is part of an open tropical forest biome, with the occurrence of three different types of subforests: open tropical forest with bamboo, open tropical forest with palm trees, open tropical forest with *cipó* vines and some areas of dense tropical forest. The climate is hot and very humid, with an average annual temperature around 26°C. The hottest period occurs in September, October and November, with average maxima of 38°C, and the coldest period is in June, July and August, with average minima of 20°C. Average annual rainfall is around 2 200 mm. The heaviest rainfall occurs in the months of January, February and March, totalling up to 800 mm. The driest period is recorded as June, July and August, with a cumulative average of up to 150 mm.

According to the RADAMBRASIL classification, the relief of the area is of the western Amazonian low plateau type, with the depression of the Acre and Javari rivers, featuring a regular platform with no rock formations such as mountains or hills. The altitude varies between 100 and 200 m.

In general, Acre soils are predominantly very fertile with a good capacity for cationic exchange, although there are patches of poor soils, which produce well in the first year, on the basis of nutrients accumulated by the ashes of the burning, but which are rapidly exhausted owing to the low technological standard of maintenance, allowing soil depletion and leaching due to high rainfall. These types of soil occur mainly in some areas of the Acre valley.

According to RADAMBRASIL, the main types of soils occurring in the state of Acre are:

- — eutrophic red yellow podzolic
- — alic red green podzolic
- — eutrophic cambissolo
- — dystrophic cambissolo
- — dystrophic gleyed hydromorphic
- — eutrophic gleyed hydromorphic
- — dystrophic red yellow latossolo.

Historical aspects

According to the National Council of Rubber Tappers, the proposal for the extractive reserve originated in the rubber tappers' struggle for identity, after being exploited for generations by the rubber barons of Amazonia. The struggle began in Xapuri, in the state of Acre, where Chico Mendes stood out as a leader in the Rural Workers' Union . This was a new event in the lives of the rubber tappers, spearheading a direct confrontation with the landowners, and gaining recognition in 1985, when the first national meeting of rubber tappers was held in Brasilia, Federal District, with the participation of 130 rubber tappers from Acre, Rondônia, Amazonas and Pará.

The movement gained momentum, on the basis of the experience gained by the rubber tappers, who were opposed to the Federal Government's development model for the region in the 1970s. The Federal Government's programme, conceived from the top downwards, included the introduction of forestry, mining, timber and farming projects, which gave rise to violent conflicts, deaths, large-scale land concentration, an exodus of traditional residents and the devastation of the region.

This movement of resistance, of struggle for land in Acre, took on a broad dimension, reflecting different, closely linked currents:

- — *Ecological:* without the forest there is no extractivism and without extractivism the land is of no use to the rubber tappers;
- — *Economic:* remaining on the land means ensuring survival;
- — *Socio-cultural:* resisting also means asserting the right to be an extractivist.

There were many specific features to this struggle which deserve consideration, one of them being the agrarian and land question. The rubber tappers were already living in the forest, from which they drew their subsistence. In the 1970s, their way of living was undergoing considerable change, especially as a result of the native rubber crisis; many of the rubber plantations were sold to southern businessmen, who turned the forest into pastureland, thus removing the means of living of the rubber tappers and nut gatherers.

The latter began to resist these changes and their resulting expulsion, combining their efforts in rural unions and organizing the so-called *"empates"* (stalemates - a form of organized, peaceful struggle to stop forest clearance).

This was the start of one of the attempts at gaining independence by the rubber tappers of the region of the Acre valley, covering the municipalities of Xapuri, Brasiléia, Rio Branco, Assis Brasil and part of Sena Madureira. This fact is fundamental for the purpose of establishing any comparisons with other forms of struggle and resistance for the possession of land in the country and may provide a basis for proposed agrarian reform in the region.

Living in the forest required a type of occupation which respected the natural distribution of species and which enabled the extractive worker to assume the cultural heritage of which he was a part. The proposal gave rise to the Projetos de Assentamento Extrativista (Extractive Settlement Projects, known as PAEs) set out in Portaria INCRA/P/No. 627 of

30 July 1987. The model of occupation met the requirements of the extractive communities as far as the spatial distribution was concerned, but, owing to its agrarian reform aspects, it needed time to come into effect. Once the advantages and drawbacks of the model were compared, the new "extractive reserve" model was chosen, based on the twin components of people and nature, to ensure that both were preserved.

In this extractive reserve model, the land belongs to the State, with the usufruct thereof going to those who work and live there.

In 1990, the aims of the land struggle were achieved. The National Council of Rubber Tappers obtained legal recognition by the Federal Government to the effect that the proposed area for the reserve was of social interest, with the result that the Chico Mendes extractive reserve was finally set up by Decree No. 99.144, on 12 March 1990.

Socio-economic data

a) Population

The Chico Mendes extractive reserve has a population of 1 838 families, totalling 12 017 inhabitants, mostly consisting of young people, aged between 9 and 17. Population density is 1.2 inhabitants per km^2. The family structure is important and decisive as regards the social division of work. There are found to be more married people than single persons or widows and widowers. Families consist on average of seven persons, with more women than men. The rubber tapper traditionally tries to keep his family together.

A natural migratory process occurs within the reserve area. Many families move from one rubber area to another looking for better living conditions. Factors conducive to this type of migration include the low productivity of the rubber growing areas, poor hunting, conflicts with neighbours, families, farmers or traders, the death of a relative and/or the removal of one of the family members to another rubber growing area.

Migration also takes place in the other direction from the town to the reserve, when the desired conditions are not met with, such as employment, schooling for children, housing, etc. There are other migration factors, in addition to the above; for instance, Brazilian rubber tappers living on the Bolivian border return to Brazil for a great variety of reasons, ranging from disappointment with the price of rubber to exploitation by traders and bosses, or just discrimination against foreigners.

Population concentration in the area will depend on geographic conditions. A rubber tapper living close to rivers, roads, a village or a municipality will have better outlets for his production and better transport; in most cases he is able to study. A rubber tapper living several days' journey away, in the more central area, practically never goes to town, depends exclusively on the local traders and undoubtedly experiences worse living conditions.

b) Production

The reserve has 1 144 active settlements or allotments (*colocações*) and 39 inactive. The average area of each settlement is 6762 hectares and each rubber stand covers an area of about 100 hectares, containing some 100 to 150 rubber trees, with an annual output of around 2.1 thousand tonnes. In these areas, with few exceptions, the Brazil-nut tree (Bertolletia excelsa), the second most important product of the productive unit (*colocaçõo*), occurs with an average concentration of up to 257 trees, of which over 50% are in full production, producing an annual average of 0.94 cans/nut tree, for a total of 340 000 cans (*latas*), amounting to 3 740 tonnes. This amount of production is considered low, due to the rudimentary type of gathering, storage and transport.

In terms of their contribution to the average liquid revenue of the extractivist (market products), rubber accounts for 44.7%, nuts for 24.6%, food 21.0%, the rearing of small animals 8.5% and other extractive products, such as fruit, oils, resins and palmitos 1.2% (CNS, 1992).

Compared with the other reserves, average output per family is among the highest, at 714 kg of rubber and 113 cans of nuts per year.

The settlements (*colocações*) are in very poor condition, as shown by the lack of any new improvements, repairs to houses, fences, meal huts, storage huts, etc.

Activities related to rubber tapping and nut gathering provide for the rubber tappers' subsistence and link the community to national and international markets. These activities are mainly the responsibility of men, although women are also involved.

The main activities in the reserve area are of the extractive type, bringing in 62% of revenues, agriculture, which contributes 29% and stock breeding, which produces only 9%. In the Chico Mendes reserve, extractivism reaches a tally of 70%, compared with other reserves. Rubber extraction absorbs around 17% of all settlement labour. Out of this total, 88% of operations are carried out by men, the participation of women being limited to about 12%.

Animal breeding plays the part of a kind of "reserve of resources" for emergencies, such as illnesses or poor hunting. This activity, which according to the general area pattern contributes 9% of income, is concentrated chiefly on cattle (4.2%), followed by pigs (3.0%) and poultry (1.8%). The breeding system used is extensive, with animals sharing the same habitat as humans, mainly poultry and pigs. Cattle, which normally occupy an average area of 2.6 hectares of pasture, are occasionally kept in corrals and shelters.

Agriculture is practised in itinerant fashion and basically at subsistence level; crops include rice, maize, beans (*feijão*) and manioc, the same area being used for not more than four years at a time.

Marketing and supplies are problematic in the reserve, owing to the distances involved and conditions of access to the settlements, as basically only the rivers can be used as access routes. The main supplier of merchandise to the rubber tapper is the *marreteiro* (trader). The produce is generally speaking sold on the spot, either for cash or in exchange for products of prime necessity.

c) Housing

Improvements such as a meal hut, small mills, presses, chicken runs, small enclosures and pig-pens, may be found in settlements with better living conditions.

The rubber tappers' residence is generally built of *paxiúba* (Iriartea sp) slats for the walls and floors, with a cover of wood strips or various types of palm leaves. The number of rooms usually does not depend on the size of the family, and consists of 3 divisions (bedroom, living room and kitchen), with a great variation in size, covering an area of between 20 up to 100 m^2.

The areas and boundaries of a settlement are always natural, marked out by rivers, waterways, lines of rubber trees or nut trees, with very few fences or boundary marks. Fences are few in number and used only to protect domestic animals. Paths and trails belong to everyone, even when they pass through the backyard of a house, and they are generally kept clean with a collective system of work (*mutirão*).

d) Education

The Chico Mendes reserve has 86 schools attended by the rubber tappers' families. Out of this total, 11 schools, with 16 teachers and 319 pupils enrolled, are financed by the Rubber Tapper Project (*Projeto Seringueiro*), which is aimed at orienting and improving literacy studies, including reading, writing and general knowledge. The state government, through its education secretariat, maintains an agreement with the Centro de Trabalhadores da Amazônia (CTA), whereby some of the teachers are paid under the agreement and some are paid by the communities.

e) Health

"Health" assistance in the area is provided by the rubber tappers themselves, who are trained to act as health agents in the settlements in which they live, where their health activities include prevention, guidance and education, with a first aid, vaccination and serum service. This working approach is part of the rubber tapper project.

Social management and organization

The rubber tappers started to be organized, in a political sense, only when the trade unions appeared in the region of the Acre-Purus valley in the 1970s. Before the unions came, there was no body as such which could channel the claims of the extractive communities.

The dominant features of life in the rubber growing areas were isolation and individual production. It was the struggle for land, in the 1970s, which profoundly affected the political life of the rubber tappers and gave them an identity and rights, from which they are now benefiting.

With their various forms of organization, the rubber tappers now possess one of the basic ingredients for administering and operating the reserves. Their participation in social and political activities has been increasing rapidly. They belong to trade unions, cooperatives, associations, political parties, church groups and community centres. The greatest participation is by men, even in religious activities such as church groups and community centres.

For the National Council of Rubber Tappers and for the National Centre for the Sustained Development of Traditional Communities (CNPT), organization inside the reserve has become a priority. This is because it is only through collective and democratic participation in different ways that the rubber tappers will be able to gain sufficient awareness and responsibility to develop their potential. Organization provides the means of establishing greater participation and greater involvement by the communities and of arriving at alternative proposals for fulfilling their aspirations.

Utilization plan

The Chico Mendes extractive reserve is defined in its utilization plan. The proposal submitted by the Association of Rubber Tappers and Farmers of the Alto Juruá Extractive Reserve to IBAMA, according to the Vice-President of the National Council of Rubber Tappers, has already been discussed with the residents of the Chico Mendes reserve, but as the local associations are not yet legally established, the procedure will be deferred until the end of 1995.

Current projects

A number of projects are currently underway to support the Chico Mendes extractive reserve. The main one of these is to include it in the Pilot Programme for the Protection of Brazilian Tropical Forests (Extractive Reserves Project), the general objective of which is to test appropriate models of economic, social and environmental management, with a view to improving the methods and procedures of traditional communities, by:

i) strengthening the organization of production and of the community;
ii) developing technologies to produce subsistence goods and products for market;
iii) developing the economic and social infrastructure;

iv) maintaining stocks of natural resources;
v) preserving biodiversity.

Current projects also include:

— *Support for community organization*

Organizing the community has always been a challenge, since man's isolation in the forest is the main characteristic of extractivism. Communities began to have a say in government policy-making in the 1970s and this participation has increased ever since. In the specific case of extractive reserves, their viability must depend on organization, since a condition of the establishment of these areas is precisely joint management between the government and an organized civil body.

— *Training to improve rubber quality*

In order to increase the rubber tappers' income, different kinds of training are being offered to improve rubber quality. The aim of the training is to introduce technology for producing smoked raw sheet rubber instead of the traditional system of producing rubber in balls.

— *Training for assistant supervisors*

Because the extractive reserve is a direct use conservation area, which means that the community is involved in its management, training is now being offered on technical, legal and administrative aspects of supervising the area. This procedure is in accordance with Brazilian environmental legislation, under the terms of the National Council for the Environment's Resolution No. 003, of 16 March 1988.

Main problems

The problems in the reserve area are innumerable. As far as health is concerned, there is a lack of medicines, delays in the vaccination schedules, no improvements in medical care, a continuing need to enlarge many of the dispensaries, and a need to improve the standards of agents by working with the schools and by encouraging the community to become more involved.

According to the socio-economic survey data of April 1992, sanitary conditions are precarious. Almost 80% of families have no cesspit. The water used comes mostly from small waterways or water holes, actual wells being few and far between. Less than 20% of families either filter or boil their drinking water.

In the education area, there is a lack of teaching material and of communication between teachers and parents. More schools need to be built and more training needs to be provided for teachers. The school curriculum should be biased towards agricultural activities, etc. There is also the underlying need for school and community organization.

In the transport sector, more branch lines need to be opened and more boats and pack animals are required. There is no pricing policy for extractive forest products, such as rubber and nuts, and production needs to be better organized with incentives, particularly in the form of farm credits.

Additional socio-economic data

Production, prices and average gross annual income per family				
Chico Mendes reserve	Production	Price	Gross income	
	KG	US$	US$	Participation %
Agriculture			683.00	47.12
Rubber	914	0.60	428.00	29.56
Nuts	1,250	0.10	125.00	8.62
Domestic animals	264	0.49	129.36	8.92
Hunting and Fishing	182	0.46	83.72	5.78
TOTAL			1,449.48	100

Source: World Bank - IBAMA/CNPT - ECOTEC: Estudo de Pré-investimento do Projeto RESEX, 1994

Chico Mendes Extractive Reserve: extraction of *copaiba* (Copaifera sp) oil *(photograph: Miguel von Behr).*

Rubber tapper's production schedule - Chico Mendes Extractive Reserve

Period / Months of activity	Rainy				Dry					Rainy		
	Jan	Feb	Mar	Apr	May	Jun	Jul	Aug	Sep	Oct	Nov	Dec
Extractive activities:												
— Latex tapping and coagulation				X	X	X	X	X	X	X	X	
— Rubber tree rest period	X	X	X									
— Plantation clearing			X									
— Nut gathering	X	X										X
Agriculture:												
— Soil preparation					X	X	X	X				
Planting:												
— Rice									X	X		
— Maize									X	X		
— Beans *(feijão)*			X	X	X							
— Manioc									X	X	X	
Harvest:												
— Maize		X	X	X								
— Rice	X	X	X									
— Beans *(Feijão)*							X	X				X
Breeding:												
— Small animals	X	X	X	X	X	X	X	X	X	X	X	X
— Dairy	X	X	X	X	X	X	X	X	X	X	X	X
Hunting:	X	X	X	X	X	X	X	X	X	X	X	X
Fishing:				X	X	X	X	X				
Marketing:	X	X	X	X	X	X	X	X	X	X	X	X

NB: The extraction of other species (e.g. medicinal, oleaginous, etc.) continues throughout the year, depending on the period of production of each species.
Source: CNS/Levantamento Sócio-econômico das Reservas do Vale do Acre/Purus. 1992.

MARINE EXTRACTIVE RESERVE OF PIRAJUBAÉ

Miguel von Behr

Geographic and environmental data

This is the first extractive reserve to be established outside the limits of legal Amazonia and also the first marine reserve. The area is part of one of the most extensive mangrove flats of the south of Brazil.

The reserve, totalling 1 444 hectares, includes 700 hectares of marine bay and 744 hectares of mangrove. It is located within the urban area of the town of Florianópolis, in the southern state of Santa Catarina, which has around 1 million inhabitants, and constitutes the largest area of mangrove trees which are still preserved on the island of Santa Catarina. As such, the reserve is essential for the maintenance of fisheries in the southern bay of Florianópolis, because the mangrove flats provide a habitat and a breeding ground for the main economically valuable marine species of the region (fishes, crustaceans and molluscs).

The marine environment of the reserve, which is made up of the mangrove flats and shoals (*baixio*, a sandbank in shallow water) is energetically interrelated through the export of organic material from the mangroves to the sea and also by the use of the mangrove trees by typical species living in the shoals and bay waters. Apart from species living among the mangroves, a large number of visiting species enter the area occasionally in their juvenile stages and use it as a breeding ground.

On the basis of existing studies and experiments, coordinated by IBAMA with the help of fishermen, the reserve was set up for the purpose of working out a proposal for managing and exploiting one of the marine species in the area, a type of bivalve mollusc known locally as the *"berbigão" (Anomalocardia brasiliana).* The species is known in the country under other names, such as the *"sarnambi"* (in Bahia), *vongôi* (in Sao Paulo), etc. This type of mollusc breeds in the shoals adjacent to the mangrove flats.

Other species of marine fauna of economic value occur in the reserve, such as the pink and white prawn, fishes, which include the mullet (*tainha* or *parati*), small cods (*pescadinha*), croaker (*corvina*), catfish (*bagre*), and some crustaceans, such as crab and shellfish.

The post-larval shrimps move among the mangroves, where they remain for four or five months, later continuing to migrate into the shoals and into the open sea, to complete their life cycle, in which the mangrove plays a fundamental role for the reproduction of the species.

The reserve area belongs to the Federal Government (mangrove flats and shoals). In this sense, the legalization of the reserve from the land point of view has already been completed, since no cost was incurred to expropriate the land or to compensate for any improvements.

Owing to its environmental value for the bay, a great part of the reserve's mangrove area had to be marked off (by means of canals) before the reserve was set up, in order to prevent the area from being destroyed as a result of the intense urbanization currently going on around the reserve.

History of the reserve

The reserve was established by Decree No. 533 of 20 May 1992.

In 1988, the SUDEPE (Superintendência do Desenvolvimento da Pesca, or Fisheries

Development Office), now part of IBAMA, seeing the need to regulate the exploitation of *berbigão* shellfish, which were being withdrawn haphazardly, with a corresponding risk for their reproduction, undertook a scientific study for the experimental establishment of a marine *berbigão* farm in the form of a pilot project. The project was coordinated by the marine biologist Ernesto Tremel. The cooperation of the fishermen was also sought, with a view to regulating the exploitation of the very large quantities of *berbigão* molluscs, which were being extracted from the area without any planning and with a serious risk of jeopardizing the ecological equilibrium of the area.

In 1992, in order to guarantee access to the area, to enable the project to continue and to obtain the necessary financial and technical support, the fishermen, backed by the IBAMA office in Santa Catarina, asked the CNPT for a study to be made of the possibility of transforming part of the bay and the adjacent mangrove flats into an extractive reserve, on the grounds that the area lent itself to the collective use of natural resources.

A number of governmental and non-governmental organizations joined in and supported the establishment of the reserve:

— Procuradoria Geral da República (General Procurator's Office);
— Movimento Nacional dos Pescadores (MONAPE) (National Movement of Fishermen);
— Colônia dos Pescadores Z11 (Z11 Fishermen's Colony);
— National Council of Rubber Tappers;
— IBAMA office in Santa Catarina;
— Environment Foundation of Santa Catarina (FATMA);
— Empresa de Pesquisa Agropecuária e Difusão de Tecnologia de Santa Catarina (EPAGRI) (Stockbreeding Research and Technology Dissemination Office of Santa Catarina);
— Ministry of Aeronautics - Florianópolis airbase;
— Delegacia do Patrimonio da União (DPU/SC) (National Heritage Office);
— Florianópolis port authorities;
— Federal University of Santa Catarina (Department of Anthropology and Biology);
— Instituto de Estudos da Amazônia e Ambientais (IEAA) (Institute for Amazonian and Environmental Studies);
— Association of Residents of the Pirajubaé coast.

Socio-economic data

Out of the 115 families which depend on the marine resources of the reserve, about 15 families, with IBAMA's authorization, extract *berbigão,* while approximately 100 exploit other marine resources of the reserve, such as fish, shrimps, oysters, crabs and other molluscs, in such a way as to guarantee the conservation of these resources.

The beneficiaries of the reserve live next to it, 5 km outside the centre of Florianópolis, in the Costeira do Pirajubaé neighbourhood, a densely populated urban area. This means that all the public services of a large town are available, such as shops, schools, leisure amenities, dispensaries, supermarkets, etc. It should be emphasised that, despite the high population density close to the reserve, the number of fishermen actually extracting *berbigão* is small in order to guarantee the reproduction of the species in the area.

Fishing is the main source of income of the reserve's beneficiaries, although 40% of them engage in secondary activities, as construction workers or in the shopping areas close to the reserve.

All extract marine resources for their subsistence from the mangroves, the sandbank and the bay itself.

Almost a quarter of reserve dependants derive income from some other economic activity, such as carpentering, sewing or by working in shops or building sites.

To catch shrimp and fish, many of the fishermen use nets. Some 40% also use the circular casting net (*tarrafa*) exclusively to fish for shrimp, for their own consumption and for marketing.

In 1990, 153 tons of *berbigão* shellfish with their shells were marketed, and 178 tons in 1991. The shellfish are extracted twice a week. A suitable annual production is estimated to be 200 tons, subject to maintaining rotation, minimum size and repopulation controls, as at present.

As the *berbigão* filters water to extract its own food, the catch is marketed and distributed after the animals have been cleaned by a specialized company, approved by the Ministry of Agriculture. There is also a sanitary check on the water contained in the product. Only then is the product sent to seafood and pasta restaurants in São Paulo and Rio de Janeiro, which have a taste for the *vôngoli*, either on its own or as part of another dish, such as the popular spaghetti with *"vôngoli"*. The *berbigão* has 30% more protein than mussels (*mexilhão*), another seafood which is much appreciated in Brazil.

Management and social organization of the reserve

A small company was set up in the reserve to market the produce of the extractive workers. An authorization for the extraction of *berbigão* is issued by IBAMA every six months, on the basis of the volume of production, for which data are collected and supplied by the fishermen to the trader under IBAMA's supervision. This is to ensure sound management of the reserve's marine resources.

The families benefiting from the reserve have accumulated knowledge and considerable experience concerning the handling of marine resources. Their tradition has given rise to the development of some highly specific typical fishing tools and practices for the capture and handling of natural resources. Preserving the cultural traditions of small-scale fishing, in accordance with the wishes of the local population, is now a priority both for the social scientists and for environmentalists.

The marine extractive reserve of Pirajubaé, as both an area and a legal instrument, is intended as a means of preserving these traditions. It is an essential part of the effort to introduce viable practical measures in support of the sustainable extraction of natural marine resources.

It is worth highlighting the work carried out by the researcher Ernesto Tremel, who carried out his scientific research alongside the fishermen, giving them the right awareness and changing their predatory activities into a rational form of exploiting the marine resources.

The type of extractivism applied in the reserve depends on integrating the three areas of mangrove, sandbank and bay fishing. It represents a significant component of the local economy and cultural tradition. The reserve guarantees the continued integration of traditional communities in the environment.

Utilization Plan

The *berbigão* shellfish are caught in accordance with a set of scientific criteria, in order to ensure that the existing population of molluscs is preserved indefinitely. In recent years, existing stocks on the sandbanks of the bay were threatened by disorderly capture, even of small molluscs, which had not yet had time to reproduce.

The solution which was found to overcome this problem was a type of equipment used in extraction similar to a rake (*gancho*), which acts like a plough, leaving behind the younger individuals which have not yet reproduced. Limits are also set both on the number of fishermen and on the volume of production.

Alongside the above measures, a scientific study was made to monitor the catches and to keep a statistical record of the average dimensions of each load, a work undertaken by Dr. Ernesto Tremel, which had never yet been done in Brazil. The "rake" operates the selection. Molluscs measuring less than 17 cm slip through the grid, together with impurities. The rakes, which are soon to become personal, in order to facilitate extraction controls, are of a simple design and manufactured by the workers themselves from soldered strips of iron. Apart from those, the only other rustic tools used are 18 kg oil cans, used as a measure, and a few sacks.

The tide is an all-important working factor for the extractivists, who have to fill the quota allowed by the utilization plan before the water rises.

Briefly, the extractive system is based on three points: selected capture, sparing the youngest specimens; rotation in the fishing grounds and continuous repopulation.

Current projects

The CNPT is currently supporting a number of measures for the effective establishment of the reserve, the last two of which have been undertaken in conjunction with the state authority supporting the development of fisheries:

1. Marking off the remainder of the mangrove flats with channels and the sandbank or shoals where the *berbigões* are located;
2. Fisheries support to encourage a better administrative and political organization of the extractive reserve's Fishermens' Association;
3. Detailed socio-economic census and survey of both professional and subsistence fishermen living around the reserve and operating within it.

Main problems

1. Invasion of the reserve area by fishing boats from other regions.
2. Increase in the numbers of amateur fishermen engaging in small-scale fishing.
3. Spread of urbanization in mangrove areas of the reserve which have not yet been marked out.
4. Lack of better organization among fishermen that do not extract *berbigões*.

Extraction of *bergigão (Anomalocardia brasiliana)* *(photograph: Miguel von Behr).*

FREXAL EXTRACTIVE RESERVE

Miguel von Behr

Geographic and environmental data

The reserve is situated in the municipality of Mirinzal in the state of Maranhão. It was established by Decree No. 536 of 20 May 1992 and covers an area of 9 542 hectares.

Compared with the reserves of Maranhão and Tocantins, the Frexal reserve offers better conditions of access. The reserve is cut through one third of the way from the west by the paved highway MA-006, running north to south and connecting the area with the town of São Luiz, the capital of the state of Maranhão, some 300 km away. It also has a few more passable roads and some internal trails.

By its location, the area belongs to the homogeneous micro-region of the Baixada Ocidental Maranhense and is included within the environmental Protected Area of the Baixada Maranhense, set up by state decree in June 1991.

The climate of the region is sub-humid and the soils are predominantly indiscriminate hydromorphic. According to the Radam-Brasil Project (a plan to map all the natural resources of Brazil), the reserve is located in the sub-coastal area of Bacabal, which is made up of a series of gentle hills, modelled on sedimentary rocks of the Barreiras and Itapecurú formations, and removed from direct tidal influence.

The predominant type of forest cover is latifoliate secondary forest, with devastated areas which induce a proliferation of *babaçu* palms, and ciliate forest along the waterways. The reserve's drainage system belongs to the Uru river basin, which contains areas of floodable fields, typical of the Baixada Ocidental Maranhense.

Two types of human activities may be easily identified: subsistence agriculture (small cleared plots where maize, manioc and rice are grown) and large-scale farming (pastureland for cattle and coffee, *guaraná* and black pepper (*pimenta do reino*) plantations).

These activities mostly take place in the higher parts of the reserve, while in the lower parts, the *buriti* (Mauritia palm) and *açai* (Euterpe palm) groves of the Uru river and other waterways are practically intact. The floodable fields are used basically to plant rice, for the extraction of fish and for extractive activities such as gathering medicinal plants.

Great quantities of *babaçu, tucum* and *pindo* plants have proliferated in abandoned deforested areas. Their current growth shows that the devastation of the original forest (pre-Amazonian) occurred several decades ago. Even the forest area known as "Mata do Jacund-87-", which shows pre-Amazonian tropical forest characteristics, has some features which are typical of secondary forest. The *pindo* groves predominate in many areas of the reserve, especially those most intensely used for cattle.

Nevertheless, chiefly along the waterways, river banks, creeks and marshy areas, features of the original vegetation are still to be found, with a predominance of *buriti, juçara* and *andiroba,* which shows that the ciliate vegetation and therefore the water sources themselves have been maintained and relatively well preserved.

History of the reserve

Historically speaking, the establishment of this reserve is connected to the long struggle of the negro community of Frexal to obtain recognition by the state of their right to land, as actually guaranteed for the remaining members of the *quilombo* (run-away slave) communities in the Brazilian Constitution. By declaring the area to be an extractive reserve, the Brazilian State recognizes the historic value of the Frexal community, as part of the nation's civilization.

It should be remembered that, before the landowners came, there was always a black community living in the area in perfect harmony with the natural environment, on the basis of stable production systems.

The area was bought in 1974 by Mr. Tomas de Mello Cruz from Mr. Adan Dietrich von Bolow, by public deed of purchase and sale, recorded in the registry of the first office of the district of Guimarães, book No. 21, folios 30 (verso) to 32.

A number of governmental and non-governmental organizations supported the establishment of the reserve and took part in the process:

- Mirinzal Workers' Union;
- Maranhense Society for the Defence of Human Rights;
- Negro Culture Centre;
- Cáritas Brasileira;
- Mirinzal Parish, attached to the Pinheiro Diocese;
- General Procurator's Office;
- State Secretariat for the Environment and Tourism of Maranhão.

Socio-economic data

About 180 families live within the reserve, split into three small communities (Frexal, Rumo and Deserto).

Although many *babaçu*-growing areas have been devastated, there are still large areas which can be exploited by the extractivists. The low prices currently fetched by the *babaçu* nuts in the market at present, however, tend to discourage local communities from gathering them, the men often preferring to work as day labourers on the farms.

According to the mapping survey of the potential of *babaçu* occurrences, carried out in 1982 by the Ministry of Industry and Commerce, the extractive reserve is part of a general 338 700 hectare *babaçu* growing area, with 40% coverage, which would correspond to an actual area of 135 480 hectares. The concentration of adult plants is as much as 74 palm trees per hectare and on average 24 productive palm trees occur in each hectare, i.e. 32.0%. The adult palm concentration refers to the average number of adult palm trees detected per module within the area of growth. Average productivity is 608.7 kg per hectare and the *babaçu* patch where the reserve is located produces 82,466.7 tons.

Traditionally, the *babaçu's* kernel is used to produce oil and milk, its leaves in basket weaving and house building and its shell to produce charcoal.

Other types of palm tree occurring in the region, such as *buriti, juçara* (or *açaí*), *tucum* or *bacaba* are used basically as food of considerable importance in the household economy. A number of "plantas ervas" (the local term, literally "herb plants") play an important part locally in popular medicine. The species most used include *contra-erva, hortelã-de-folha grossa* (mint), *mastruz, arruda, boldo, quina, pega-pinto, para-tudo, algodão* (cotton), *jataúba, janaúba, bacuri, santa-maria, araticum, catinga-de-bode, junca, goiaba-azeda, andiroba, maria-cagona, quando-é, assussena, juçara (root), pinhão branco* and *jaborandi.*

Some species of *cipó* are also quite frequently used, such as *titarra*, used to make baskets for fishing, and *cipó-branco* and *curuapé,* used to tie fence posts and for fastenings in house building.

Several species of fauna occur in the region, such as: paca, deer, *capivara, cutia* (agouti), *caititu* (peccary) and others. However, owing to the strong pressures generated by deforestation and subsistence hunting, the populations of these species have been diminishing considerably. In fishing, the most consumed species are the *traíra, pacu, aracú, piranha, piau, piaba,* etc.

As far as education is concerned, there are only basic literacy schools in the reserve villages, which do not complete the first grade. There are some dispensaries, but which operate under precarious conditions.

Management and social organization of the reserve

The reserve has three residents' associations: Frexal, Rumo and Deserto, which are still not fully integrated from a management point of view. A considerable effort is being made at present by the community's residents' association, which bears the same name as the reserve (Frexal) to set up a single association, which, according to the residents, will emerge more naturally once IBAMA has obtained the land rights over the area.

Utilization Plan

In the reserve, extractive plant activities, hunting, fishing and subsistence agriculture coexisted in harmony with the environment before the latter suffered the impact of external agents, in the event that of the last great landowners. The type of multiple use of the reserve's natural resources, which is currently being developed by the residents on what has been left after the devastation, offers a production model which has the advantage of exerting least pressure on the land.

The harvesting of the *babaçu* palms, even though they are not being exploited very intensively at present, could be improved in terms of productivity by thinning out the palm trees which are found to be unproductive and by controlling the overpopulation of *pindova* (little developed *babaçu* palms) and by combining *babaçu* forests with perennial cultures adapted to the region (May 1990).

The community is hoping that an agricultural college will be set up in some large free premises close to the Frexal community, which would be a way of giving greater effect to discussions and would assist projects in favour of the sustained development of the reserve's natural resources. The characteristics of the Frexal Quilombo extractive reserve are ideal for such projects.

Considering the great expanses of damaged areas in the territory of the reserve, its development plan should lay special emphasis on recovering degraded areas, as a test experiment for the whole region of the Baixada Maranhense. The reserve could then act as a pilot unit for experimenting with the sustainable use of the natural resources on which traditional communities live.

Current projects

The land survey of the reserve was recently completed (the official registry survey and evaluation of lands and improvements), opening the way for IBAMA to expropriate the reserve lands. Thanks to the survey, which was well supported by INCRA (Instituto Nacional de Colonização e Reforma Agrária), it was possible to arrive at an estimate of the cost of expropriation.

Main problems

1. Lack of better community organization among the three local associations;
2. Because of its specific cultural, social, ecological and geographic characteristics, the measures taken for the reserve should be completely different from those applied in other reserves;
3. Lack of any specific policy for products derived from the extraction of *babaçu,* for instance through the opening of credit lines suited to the cultural, ecological, economic and social conditions of the extractive communities.

The *buriti* and *açaí* groves along the banks of the Uru river and other waterways are still practically intact *(photograph: Julio Ruiz Murrieta).*

RIO CAJARÍ EXTRACTIVE RESERVE

Adalberto Iannuzzi Alves

Historical aspects

The Rio Cajarí extractive reserve is located in the southern part of the state of Amapá, a region which was occupied during the rubber cycle halfway through the 19th century. The land was initially appropriated by the rubber planter José Júlio de Andrade, who, during his terms of office as federal deputy and senator, managed as a result of back-office manoeuvring to register an area in excess of 3 million hectares, including lands in the states of Pará and Amapá.

In that period, the region attracted a significant number of people from the north-east and peasants, who engaged only in extractive activities, such as the gathering of rubber, *balata, maçaranduba,* Brazil nuts, wild animal skins, medicinal oils, etc. Meanwhile, Colonel José Júlio, as he was known in the region, was accumulating riches by exploiting the work of these people.

After the political decline of the colonels, in 1948, the lands were sold to a group of Portuguese businessmen, who besides setting up a trade and navigation company, maintained their hold over the extractive workers through the *aviamento* debt peonage system. This company stayed in the region for 19 years and then sold off the "property" to the American millionaire Daniel Ludwig, the founder of the Jarí project.

The basic objectives of the Jarí project were to introduce forest activities (forestry and manufacture of cellulose) and farming (rice production and the rearing of bubaline animals), in addition to mining (kaolin and bauxite) and logging.

When the Jarí project was launched, traditional extractivism was abandoned and the company began to undertake large-scale deforestation in nut and rubber growing areas, for the purpose of planting exotic forest species to produce cellulose.

Without the support of employers, part of the communities abandoned extractive activities for small jobs on the Jarí project. Others left the region and those who remained were obliged to submit to the economic exploitation of the *"regatões"* (itinerant traders), who, with the support of Brazil nut and rubber processing industries of the state of Pará, began to supply the local population through the system of *aviamento.*

When the Jarí company almost went bankrupt, in 1981, a group of Brazilian businessmen, financed by the government, took over the company's activities. Since then, the new Jarí company has been encountering many problems in its current affairs, due mainly to the low fertility of soils, erosion, the inability of some species to adapt to the hot spells, the emergence of plagues and diseases, amongst others.

In order to protect such a huge area of land, the Jarí company set up its own "security" force, which started to persecute the local people, preventing them from performing their normal activities. As conflicts broke out, the extractive workers began to organize, initially setting up an association, which later became a cooperative. At the same time, discussions began about the ownership of the land and the usufruct of natural resources, under the leadership of the Rural Workers' Union of Macapá.

In 1985, at the time the National Council of Rubber Tappers was set up, the Macapá Rural Workers' Union also joined in the struggle to establish protected areas for the rubber tappers, like those set aside for the Indians. This proposal became part of the extractive settlement projects (the PAEs) of the National Agrarian Reform Programme (1987), related to extractive reserve environmental policy (1990).

In this new situation, in October 1987, work began on defining areas for the introduction of PAEs in the state of Amapá. As a result of these studies, which were coordinated by the then Ministry for Agrarian Reform and Development (MIRAD), with the participation of the Macapá Rural Workers' Union, the Cooperativa Mista Extrativista Vegetal dos Agricultores de Laranjal do Jarí (Laranjal do Jarí extractivist cooperative), the Associação de Assistência Técnica e Extensão Rural do Território Federal do Amapá (Amapá technical assistance and rural development association) and the Institute for Amazonian Studies, the extractive reserve projects of Rio Maracá I, II and III were established in 1989, and the Rio Cajarí extractive reserve in March 1990.

Geographic and environmental aspects

a) Climatic characteristics

The main characteristic of the Rio Cajarí reserve is its high rainfall, with annual precipitation varying between 2 300 and 2 400 mm. Most of the rainfall occurs in the months of February, March and April, and the driest period is September to December. According to the Koppen classification, the climate is Af in the area near the Canal do Norte (Amazon river) and Am in the centre/north of the reserve, where the Campo Cerrado vegetation predominates.

Relative air humidity averages 85% annually, and temperature varies between 16°C and 38°C throughout the year,with an average value around 26°C. Average annual sunshine varies between 2,000 and 2,200 hours.

According to the Bagnouls and Gaussen bioclimatic classification, the area is attenuated thermoxerokimenic in transition to mesoxerothermic with up to 04 biologically dry months in the centre-north.

b) Hydrographic network

The reserve is drained by the basins of the Cajarí and Ajuruxí rivers, Igarapé Tambaquí and other small waterways coming from the north-east, and all the waters flow towards the Canal do Norte, Amazon river.

In the upper and middle courses of the Cajarí river, the soils are characterized by excessive incipient drainage, and the alluvial plain favours the accumulation of water and the formation of inundated terraces, known by the residents as the *várzeas, or* flooded lowlands.

Along its lower course, the Cajarí river and other waterways intermingle with the areas of alluvial deposits of the Canal do Norte, forming meanders, dykes, channels, branches and lagoons. High and low water-marks depend on the tides.

c) Soils

The reserve's soils are mostly of medium to clayey texture, low natural fertility, with concretions and vulnerable to erosion; they belong to the following major groups:

- red yellow latossolo
- yellow latossolo
- dystrophic red yellow podzolic
- hydromorphic
- eutrophic and dystrophic gley
- eutrophic and dystrophic alluvial
- dystrophic lithoidal

d) Vegetation cover

Existing vegetation formations in the Rio Cajarí extractive reserve are as follows:

Savannah/forest contact

There are two types of transition areas (contact) with clear characteristics: (i) Campo Cerrado, which occupies 9% of the area and occurs at the head of the Cajarí river, in the centre and north-east of the area, and (ii) Parque, occupying 5% of the area, and occurring in the north-eastern part of the reserve.

Pioneer formations

Open alluvial vegetation: occupies 12% of the area and occurs along the middle and parts of the lower Cajarí.

Dense forest occurrence

Dense ciliate tropical rainforest of the low plateaux: occupies 38% of the area, the main representative forest species being the *angelim* (Dinizia excelsa), *maçaranduba* (Manilkara huberi), *faveira* (Parquia pendula), *tauarí* (Couratari spp), *tachís* (Tachigalia sp), and *jutaí-açú* (Hymenaea oblongifolia).

Dense ciliate tropical rainforest of the terraces: occupies 1% of the area, with a high occurrence of *anani* (Sinphonia globulifera), *açaí* (Euterpe oleracea), *virola* (Virola surinamensis), *ingá* (Inga sp), rubber trees (Hevea brasiliensis), and *itaúba* (Mezilaurus spp).

Dense tropical rainforest of the submontane plateaux: occupies 13% of the reserve, with the following representative vegetation: *tauarí* (Couratari spp), *faveira* (Parkia spp), *angelim* (Dinizia excelsa), *maçaranduba* (Micropholis guianensis), *louro prata* (Ocotea guianensis), and *samaúma* (Ceiba pentandra).

Dense tropical rainforest of the interriver channels: occupies 10% of the reserve, with concentrations of palm trees, mainly *babaçu* (Orbigua martiana), *ucuúba* (Virola surinamensis), *buritî* (Mauritia flexuosa) and rubber (Hevea brasiliensis).

Dense tropical rainforest of the high plateaux: occupies 12% of the area, with vegetation including maçaranduba **(Manilkara huberi),** *faveira* (Parquia pendula), *taxís* (Tachigalia sp) and *cipó* forest.

e) Geology and geomorphology

Geologically speaking, the area of the Rio Cajarí extractive reserve is situated within the Amazonian platform and consists predominantly of unconsolidated cenozoic sediments covering paleozoic sedimentary rocks.

f) Fauna

The characteristics of the reserve region's fauna differ according to the ecosystems where they occur, with different habits, feeding, etc.

In the dense forest ecosystem, there are *anta* (tapir), *cotia* (agouti), sloth, *quatí,* monkeys and cats.

In the Campos de Cerrado (mixed savannah) ecosystem, there are mainly animals of terrestrial habit, such as reptiles, small rodents, herbivorous mammals, etc.

The riparian and freshwater ecosystem sustains a more abundant fauna, both land and water based, such as alligators, turtles, otters, (lontra) and *ariranha, capivara,* tapir, *paca,* and a great quantity of birds, such as *marrecos,* heron *(garças),* socó, ducks, kingfishers (ariramba) and fish of all kinds.

Socio-economic data

a) Population

Population origin: The origins of the Cajarí river extractive reserve population are as follows:

— Amapá - 84%
— Pará - 12%
— Maranhão - 02%
— Other - 02%.

Main population characteristics

— Number of resident families: 672;
— Population density: 0.721 inhabitants/km2;
— The population is mainly young, some 78% of residents being under 30 years old;
— Family units made up of 3 to 8 persons;
— 31% of families live in small villages, while 62% live dispersed along the rivers.

Influence of the availability of natural resources on the formation of villages

— The proximity of nut and rubber trees contributed to the formation of the following localities: Água Branca, Marinho, Boa Esperança, Sta. Clara, Sororoca, Acampamento, Santarém and D. Maria.
— On the other hand, the communities of Maranta, Ajuruxí, Santana do Cajarí and São Pedro, which are located in the low-lying flood areas, are based economically on the gathering of rubber, fruit, açaí palm hearts and various woods.

b) Existing infrastructure

For better comprehension, the reserve has been divided into three different sectors, firm land, Cajarí river and Ajuruxí, where the existing infrastructures in each community are as follows:

Cajarí river sector: This sector is fairly remote from the urban centres; access is only by boat; the population is 981 persons, distributed over 174 families.

— São João - dispensary and school;
— Santa Rita - no dispensary, school operating in private house;
— Itaboca - one school;
— Santana do Cajarí - one school operating precariously;
— Conceição do Muriacá - school and dispensary;
— São João II - dispensary and school;
— Boca do Braço - one school.

Ajuruxí sector: This includes the coastal part and the lake region of the Ajuruxí river and the Ariramba. The population is 1 352 persons, distributed over 227 families.

— São Pedro - neither school nor dispensary;
— Macedônia - one school, no dispensary, but one nurse;
— São José - one school operating precariously in a private house;
— Santo André - neither dispensary nor school;
— Ariramba - school and dispensary;
— São Bernardo - school and dispensary;
— Maranata - school and dispensary.

Dry land sector (above flood level): This includes the areas of influence of highway BR 156; it has easy communications with urban centres and a population of 1 146 persons, distributed over 221 families.

- — Acampamento - one school;
- — Água Branca - school, dispensary and store under construction;
- — Marinho - school and nurse;
- — Santa Clara - school, store and store under construction;
- — Boa Esperança - no infrastructure at all;
- — Sororoca - no infrastructure;
- — Açaizal - no infrastructure;
- — Santarém - one school;
- — Dona Maria - one school.

c) Production

In order to ensure its survival, the resident population of the Rio Cajarí extractive reserve applies its labour force to extractive activities and small-scale farm production. It is organized on the basis of family production units, with all the labour force distributed according to a schedule related to forest cycles. Productive areas are not necessarily continuous, but are distributed in a random fashion, according to the physical and geographical conditions imposed by nature (such as proximity to rivers) or biological conditions (distribution of plant species in the forest). The population maintains a very close relation with nature, either in the forest or on the rivers, making use of experience and knowledge accumulated throughout many generations.

Extractive activities

These activities follow a system which has been worked out over the years by the local population, whereby nature offers a concentration of species supplying raw materials, and the inhabitants for their part abide by forest cycles, alternating extractive and farming activities. Within this system, forest areas used for gathering are referred to as *"pontas"* (point), such as Ponta de Castanhal (Nut Tree Point) or Ponta de Açaizal (Açaí Point), or *"estradas"* (trails) in the case of rubber. The points are normally considered private as far as their main activity goes, such as the groves of nut trees, although this is not so for subsistence hunting and/or the gathering of *cipó,* for instance. Similarly on the rivers, the fishing activity is free, the only restriction being the ban on predatory appliances, such as the *timbó* poison vine, explosives, etc.

The main products gathered by the population of the Cajarí river reserve are the Brazil nut, in the dry land sector, and *açaí,* either for its fruit or for its palm heart, in the Ajuruxí and Cajarí sectors. Rubber gathering, which once had its golden era in the region, is currently paralyzed owing to the low prices paid for the product, in addition to very difficult marketing conditions.

d) Marketing

Dry land sector

In this sector, all transport is done by vehicle along highway BR 156, and the production is taken up by the population of Laranjal do Jarí (62%) and Macapá (13%), the remainder staying within the reserve itself (21%).

Ajuruxí sector

All transport is by river in small boats, and the most sought after markets are the two municipalities of Santana (58%), Mazagão (5%), Macapá (1%) and the reserve itself, where 32% of the products are sold.

Cajarí river sector

As in the Ajuruxí sector, all produce is taken to market by boat, mainly to the municipalities of Laranjal do Jarí (51%), Macapá (17%), Santana (15%), while 14% is sold within the reserve itself.

Social organization

The community of the Rio Cajarí reserve is organized around the Association of Workers of the Extractive Reserve of the Cajarí River Valley (ASTEX-CA), which was founded in September 1991 with the main objective of preserving the forest with its ecosystem and guaranteeing for local populations the self-sustainable exploitation of renewable natural resources, especially extractive production, with a view to attaining ecological equilibrium and a sound quality of life for forest dwellers.

The association has four decision-making bodies: (i) the General Assembly (which meets annually), (ii) Deliberative Council (consisting of 13 members, two of whom are appointed by the CNS/Amapá), (iii) Executive Board (made up of a Chairman, Vice-Chairman, secretary and treasurer) and (iv) Supervisory Council (with three office holders and three substitutes). Despite this theoretical structure, only the Executive Board operates normally. All the association's activities for the time being are carried out in conjunction with the CNS/Amapá.

Main problems

The Cajarí river extractive reserve community, as elsewhere in Amazonia, suffers from a high rate of illiteracy, serious health problems, major transport and production outlet difficulties, deficient communications and fragile base organizations.

The workers' association, which was set up by the extractivists as a way of strengthening the community organization and fostering the harmonious development of the reserve, is still not entirely fulfilling its function, giving rise to a need for training for its members in the areas of administration, community organization, fundraising and planning.

Current projects

Several projects have been launched to help strengthen the extractive reserves. The main one has been to include four of these units in the Pilot Programme for the Protection of Brazilian Tropical Forests - (Extractive Reserves' Project), one of the units chosen being the Rio Cajarí.

The general objective of this project is to test appropriate models of economic, social and environmental management, with a view to improving the methods and procedures used by traditional communities, in the administration of renewable natural resources of the tropical forests, by joint management between the government and society. The purpose is to ensure the generation of sufficient income, social equity and the exchange of experience, all on the basis of the conservation of natural resources.

The specific objectives of the subproject are: i) strengthening the organization of production and of the community; ii) developing technology to produce subsistence goods and products for market; iii) developing the economic and social infrastructure; iv) maintaining stocks of natural resources; v) preserving biodiversity.

Other current activities:

— Training of assistant inspectors

This is an activity which is recognized in Brazilian environmental legislation and which consists in training local communities in the technical, legal and administrative aspects of inspection. The persons trained are known as "assistant inspectors" and whenever they come across any illegal activity within the reserve, under the collective *(mutirão)* system, they draw up circumstantial reports of the facts, which are then sent on to the legal body in charge of applying the law.

— Support for community organization

The main feature of extractive activity in the Amazon region is the isolation of people in the forest, the lack of communication with urban centres and the practically total autonomy as regards subsistence. Organizing these populations was always a considerable challenge.

Since the 1970s, this situation has been changing gradually and in recent years many efforts have been directed at extending basic community organization. In the specific case of the extractive reserves, such organization is fundamental, considering that one of the preconditions of the proposal is the joint management of the reserves by the government and organized civil society.

— Construction of community stores and canteens

The trading system most commonly found in the Amazon region is run by the *"regatões"* (itinerant traders), who exchange goods of prime necessity for products extracted from the forest, a form of exchange which is practically invariably unfavourable to the extractive workers, insofar as they pay high prices for the goods they buy and in exchange receive very low prices for their products.

To break this vicious circle, a number of community stores are being set up, at strategic points in the reserve, in order to do away with the old system and organize the marketing of produce.

Other socio-economic data

Occupations of heads of families before coming to the Rio Cajarí Reserve	
— Extractive activities:	42%
— Agriculture:	30%
— Urban work:	25%
— Gold mining:	3%
TOTAL:	**100%**

Extractive work experience in the Rio Cajarí Reserve	
— > 15 years:	78%
— 09-15 years:	10%
— 03 A 08 years:	5%
— Less than 3 years:	7%
TOTAL:	**100%**

Family residence time in Rio Cajarí Reserve	
— > 15 years:	78%
— 09-15 years:	10%
— 03-08 years:	5%
— < 03 years:	7%
TOTAL:	**100%**
Forms of land occupancy in Rio Cajarí Reserve	
— Spontaneous occupation:	49%
— Successin:	24%
— Empresa Jari concession:	8%
— Purchase from third parties	5%
— Government concession:	2%
— No occupancy:	12%
TOTAL:	**100%**

SOURCE: World Bank - IBAMA/CNPT- CNS. Socio-economic survey of Rio Cajarí Extractive Reserve, 1993.

Most common illness occuring in families of Rio Cajarí Reserve	
Illnesses	**Families (5)**
Respiratory problems	60%
Diahorrea	48%
Worms	48%
Malaria	33%
Accidents	12%
Measles	4%
Other	11%

Relative participation by families interviewed in active groups in Rio Cajarí Reserve	
Groups	**Participation (%)**
Base communities	66%
Trade union	24%
Association	20%
Canteen/store	16%
Cooperative	4%
Other	3%

Main problems in Rio Cajarí Reserve	
None	45%
Land conflicts	17%
Sale of allotment	2%
Access to nut trees	5%
Illegal fishing	3%
Illegal hunting	2%
Other	23%

SOURCE: World Bank - IBAMA/CNPT - CNS. Socio-economic survey of Rio Cajarí Extractive Reserve, 1993.

Action of production units in extraction of forest products - Rio Cajarí Reserve

Products	Reserve total		
	Gathering	Selling	Selling > 50 %
1. Brazil nut	38%	36%	35%
2. Açai fruit	57%	7%	4%
3. Palmito	25%	25%	25%
4. Rubber	0%	0%	0%
5. Latex	4%	4%	4%
6. Amapá	3%	1%	1%
7. Andiroba	9%	2%	1%
8. Cocoa	5%	1%	1%
9. Cipó titica	16%	3%	1%
10. Other cipó	12%	1%	1%
11. Straw	19%	1%	1%

SOURCE: World Bank - IBAMA/CNPT - CNS. Socio-economic survey of Rio Cajarí Extractive Reserve, 1993.

Table of main crops and related commercial activity - Rio Cajarí Reserve			
Crops	**Reserve total**		
	Planting	**Selling**	**Selling > 50 %**
1. Rice	22%	7%	3%
2. Maize	51%	27%	13%
3. Manioc	72%	52%	36%
4. *Macaxeira*	64%	24%	11%
5. Sweet potato	52%	19%	9%
6. *Cará*	53%	20%	9%
7. Watermelon	22%	8%	4%
8. Pineapple	57%	18%	5%
9. Avocado	33%	7%	3%
10. Banana	64%	28%	16%
11. *Cupuaçu*	49%	13%	6%
12. Orange	58%	17%	9%
13. Lemon	58%	17%	9%
14. Tangerine	46%	10%	5%

SOURCE: World Bank - IBAMA/CNPT - CNS, field research, August 1993.

Communities and families - Rio Cajarí Extractive Reserve	
Community	**Families**
ACAMPAMENTO	15
ÁGUA BRANCA	48
MARINHO	42
SANTA CLARA	19
BOA ESPERANÇA	12
SOROROCA	13
AÇAIZAL	8
SANTARÉM	13
DONA MARIA	20
SÃOO JOÃO	30
SANTA RITA	12
TABOCA	11
SANTANA	10
CONCEIÇÃO	15
SÃO JOÃO II	13
BOCA DO BRAÇO	12
SÃO PEDRO	18
MACEDONIA	11
SÃO JOSÉ	21
SANTO ANDRÉ	28
ARIRAMBA	55
SÃO BERNARDO	18
MARANATA	30

Rio Cajarí Extractive Reserve *(photograph: Miguel von Behr).*

RIO OURO PRETO EXTRACTIVE RESERVE

Adalberto Iannuzzi Alves

Historical aspects

Major changes occurred in the 1970s in the former territory of Rondônia, now a state, mainly as a result of the Federal Government's policies and strategies for the occupation of frontier areas in the north of the country.

As part of these policies, a number of colonization projects were launched, slowly changing the system of production which had been employed until then in the region, which had been based on extractive activities related to rubber, nuts and other regional products. The change occurred because the migrants who settled in the area had a different approach to production, requiring more and more deforestation to plant their crops.

In 1981, with the launch of the Polonoroeste Integrated Programme for the Development of North West Brazil, financed by the World Bank (US$ 411 million), and the resulting building and paving of highway BR 364, the westward migration increased still further, profoundly altering practically all the social, cultural and environmental structure of the state in less than 10 years, despite the undertaking in the original document to direct colonization towards an area of 410 000 km^2, between the states of Mato Grosso and Rondônia. The Polonoroeste programme also proposed to settle small farming communities practising self-sustaining agriculture, with basic services in the areas of health, education and production outlets, while protecting the forest and guaranteeing the maintenance of the lands and cultures of indigenous communities.

While it was in effect, the programme led to the highest rates of deforestation of the whole region. From 1 217 km^2 in 1975 the area grew to 30 046 km^2 in 1987 and, despite all the financial resources invested there, it was far from achieving the proposed objective, and may be considered a disaster from both the environmental and the social point of view.

Considering the circumstances, several non-governmental bodies launched a national and international campaign to halt the programme, and it was against that background that the proposal to set up an extractive reserve gained momentum in the state of Rondônia. In July 1988, responding to strong pressure from several extractivist and consultancy organizations, the government of the state of Rondônia, by decree, established an environmental policy for the rational occupation of state lands, known as ecological socio-economic zoning.

Among the six areas identified in the zoning, one of them, zone IV, became the basis for the organization and development of extractivism in the policy of the government of Rondônia. Considering, however, that the proposal did not take account of the extractivists' main claims, the zoning project was chosen by the non-governmental organizations as one of their main means of pressure to oblige the state government to introduce changes both in the proposal and in the way its policies were brought into effect.

At the same time, the Acre rubber tappers, fearing that the same problems as were occurring in Rondônia might be repeated, began a movement to oppose the paving of the BR-364 highway, on the stretch from Porto Velho to Rio Branco, in the event that no measures were taken to guarantee the rubber tappers' access to land and respect for indigenous communities.

After successfully resisting the attempts at deforestation by the landowners, through their stalemate *"empate"* approach, the rubber tappers began to organize and to extend the movement to the state of Rondônia, thus laying the foundations for the establishment of the extractive reserves.

The rubber tappers became much more active in their attempts to establish new rules for the use and occupation of land, and even more so to escape from the prevailing commercial system of debt peonage or *aviamento*. As a result of this increased mobilization, the National Council of Rubber Tappers (Conselho Nacional dos Seringueiros - CNS) was set up in 1985, and under the leadership of Chico Mendes began to seek whatever political support they could with a view to becoming recognized as the protectors of the forest.

One of the results of the movement was the first state meeting of rubber tappers and rubber fighters (Seringueiros e Soldados da Borracha) in February 1989, with the support of the National Council of Rubber Tappers (CNS), the Institute for Amazonian Studies (IEA) and the State Forestry Institute (IEF/RO), a meeting which owing to the assassination of Chico Mendes in October the year before was coordinated by his cousin, Raimundo Barros. The meeting, which was held in Guajará-Mirim, emphasized the need for rubber tappers to have protected areas for themselves like those set aside for indigenous communities. Like other similar meetings occurring in the states of Amazonas, Amapá, Pará, Mato Grosso and Acre, it always attracted large numbers of rubber tappers and sympathizers who supported the new proposal.

Then in 1989, the Instituto de Terras de Rondônia (ITERON) (Rondônia Land Institute), with the technical assistance of other state and federal institutions, carried out a socioeconomic, land, soil and flora survey of the Ouro Preto river basin, for the purpose of establishing a state extractivist forest. These studies then served as a basis for setting up the Rio Ouro Preto extractive reserve, in March 1990.

Geographic and environmental data

a) Location

The Ouro Preto river is an affluent on the left bank of the Pacás Novos river, part of the Mamoré river basin, in the municipalities of Guajará-Mirim and Vila Nova do Mamoré. The area of the reserve is situated between 64° 01' 18'' and 65° 10' 32'' longitude west and 10° 35' 11'' and 11° 02' 30'' latitude south.

b) Boundaries

North: Gleba Samaúma, Setores Bananeira, Pacaás Novos (Soldados da Borracha), Evandro da Cunha (Soldados da Borracha) of the Guajará-Mirim land project;
South: Gleba Samaúma, Setor Evandro da Cunha (Soldados da Borracha) of the Guajará-Mirim land project;
East: Urú-Eu-Wau-Wau indigenous area
West: Pacaás Novos river

c) Climate

The reserve's climate is classified as Am, its main feature being average annual rainfall of the order of 2 200 mm, with the rainy quarter in December/January/February and the driest quarter in June/July/ August. Average annual temperature is above 24°C, with an absolute annual maximum of 38°C and absolute minima which can be as low as 4-6°C, during the cold spells of July. Relative air humidity averages 85% annually, with on average 2 000 hours of sunshine. Evaporation is high as a result of the high rainfall and temperature, with an annual average of 900 mm, potential annual evaporation of 1 400 mm and real annual evaporation of 1 200 mm.

d) Plant/bioma typology

In biogeographical terms, the Rio Ouro Preto extractive reserve area is situated within the Amazonian domain, in the Madeira river region, the dominant bioma type being humid tropical forest.

In smaller proportion, there is a region of pioneer formations, and a subregion of floodable accumulation areas, with tree formations, situated in depressed areas under the alluvial influence of hydromorphic gley soils. The species of greatest economic interest in the area are the *itaúba, maçaranduba, sorva, caucho* (gumtree), *copaíba,* rubber and brazil nuts.

e) Geology

The whole area of the reserve is situated in the geological province of cenozoic deposits, which are practically entirely the products of epeirogenic movements, typically producing alluvial, eluvial and colluvial cover.

The upper pleistocene and halocene sediments are more represented in this province, always with the dominance of fluvial deposits. The former, with greater spatial representation, are made up of argillite, siltstone, interdigital, clear to red arenite; the latter restricted to current drainage, consist of gravel, sand, silt and unconsolidated clay.

The relief takes the form of a flat area, which includes quaternary sediments and subsidiarily pre-cambrian lithology, the predominant vegetation cover being open forest interspersed with savannah and in more restricted areas pioneer formations.

f) Hydrography

The area of the reserve is drained by the Ouro Preto river basin, whose main tributaries are the Concordia, Repartição and Amarelo waterways. The Ouro Preto river takes its source in the Serra do Pacaás Novos, and flows into the Pacaás Novos river not far from where the latter flows into the Mamoré river.

In winter, the Ouro Preto is navigable for boats of small and medium tonnage (15-30 tons), while in the dry period (summer), only small boats and canoes can use the river.

g) Soil

According to EMBRAPA data, the area of the Ouro Preto river extractive reserve predominantly contains associations of little humic gley soils, of a clayey texture, hygrophilous flooded lowland *(varzea)* equatorial forest phase + dystrophic alluvial soils, moderate A horizon, indiscriminate texture, non-deciduous varzea equatorial forest phase, both Tb, flat *varzea* relief.

h) Agricultural aptitude

According to surveys carried out under the RadamBrazil project, the soils of the reserve area are mostly unsuitable for intensive agricultural use, the land being more appropriate for the conservation of flora and fauna.

Socio-economic data

a) Population

The social groups living off extractive activities in Amazonia generally descend from northeastern migrants, which came to the region at two different times. The first was at the beginning of the century, when some 500 000 people came to the region, and the second time was during World War II, when fewer people migrated.

About 750 persons, in 123 families, live in the Ouro Preto river extractive reserve, depending economically on the use of existing natural resources and on subsistence agriculture. These families are spread along the water courses, which facilitate their movements inside and outside the area, and their distribution base is the *"colocação"* (settlement).

Each settlement is a productive unit, made up of about three rubber stands, with on average 150 trees per stand. The boundaries of each settlement are normally set by the rubber tappers according to some geographic feature, which does not necessarily coincide with areas covered by the rubber stands.

b) Infrastructure

The Ouro Preto river extractive reserve in Rondônia, through its association (ASGM), has a good operational infrastructure, as follows:

Transport:
- 1 Toyota jeep
- 4 Three-ton capacity boats
- 6 Aluminium boats (gliders)
- 2 Outboard engines (25 hp and 40 hp)
- 9 7 hp *rabeta* engines
- 2 Yanmar diesel engines
- 10 Pack animals

Communication:
- 2 Amateur radios

Education:
- 7 Schools

Health:
- 4 Dispensaries

c) Production

In the production area, there are two models currently used in the reserve, the extraction of forest products and subsistence agriculture.

Among the extractive activities, the main product gathered is rubber, for which production is concentrated in the region of the rivers Pacaás Novos, Ouro Preto and Novo. Most of the rubber tappers live in settlements (*colocações),* normally located along the banks of these rivers and/or affluents, precisely where there tends to be a high density of native rubber trees of the Hevea brasiliensis species. On the higher ground, in the so-called dry land forests, the predominant species is the Hevea benthamiana, which is used as an economic alternative during the rainy period, when the lowlands are flooded.

The rubber tappers of the region produce an average of 1 200 kg of rubber per family per year, of the CVP - Cernanbí Virgem Prensado type.

Another important constituent of the rubber tappers' income is the Brazil nut, the fruit of which is gathered during the rainy period in alternation with rubber, respecting the forest cycles. The nut trees are generally concentrated on the higher ground, together with the noble woods.

There is also extraction of *copaíba* oils (much used in regional medicine), patauá (used as a food), *babaçu, bacaba* (food) and *açaí* (food and palm heart).

Fishing is an important source of protein for the communities living in the areas of the middle and lower Ouro Preto river, while hunting is more abundant in the upper reaches.

d) Marketing

One of the reasons why the rubber tappers organized throughout Amazonia was to seek suspension of the payment of "rent" to the former bosses. This rent was a 15% levy paid each year on the output of rubber and nuts. Apart from paying this levy, the rubber tapper had to undertake to hand over his whole output to the boss, that is, he did not have the right to market his product himself. This was one of the major causes of the unpayable debts of the rubber tappers, since they were sold goods at very high prices and gave in exchange the products they had gathered, normally for prices well below those of the marketplace.

Another feature of the domination by the bosses was the ban on setting up any cleared plots of land for subsistence crops, on the pretext that this activity was prejudicial to the extraction of rubber, although basically the policy was aimed at increasing the rubber tapper's dependence on his boss, since it deprived him of any means of surviving outside the system of debt peonage or *aviamento*.

In the specific case of Guajará-Mirim, the effort to suspend payment of the rent took a fairly dramatic turn, bringing about an immediate reaction by the rubber owners, who blocked the whole system of *aviamento* and the marketing of produce within the reserve area.

In order to overcome these difficulties, the rubber tappers set up a municipal committee, which, with the support of the diocese of Guajará-Mirim and the National Council of Rubber Tappers, amongst others, bought a boat and some goods and began to set up an alternative system of supplies and marketing; this was the first step towards setting up the Rubber Tappers' Association of Guajará-Mirim.

Social organization

The rubber tappers of the Rio Ouro Preto extractive reserve have come together in the Rubber Tappers' Association of Guajará-Mirim (ASTM), which was founded in March 1991, with the following main objectives:

- i) to defend the socio-economic interests of its members, through mutual assistance;
- ii) to generate income enabling its members to be self-sustaining and to improve their standard of living, subject to preserving the natural resources of Amazonia;
- iii) to defend rubber and extractivist policies;
- iv) to sponsor the cultural and social improvement of its members.

The association's decisions are taken by its General Assembly, which is held every two years, and by its board, made up of a chairman, vice-chairman, first and second secretary and treasurer, and fiscal advisor. The board meets every six months. The association's operations are headed by eight members, including five board members and three contracted officials, a driver, a secretary and an accounting advisor.

Each member's responsibility is established during the planning phase, normally for the six months ahead.

Apart from its own staff, the association receives technical support from other non-governmental institutions, such as the National Council of Rubber Tappers (CNS), the Institute for Amazonian Studies (IEA), the Pastoral Land Commission (CPT), etc.

Utilization Plan

The extractive reserves are managed according to a utilization plan. This plan is established by the communities and its basic principle is respect for their way of life.

The utilization plan contains the following sections:

Purposes of the plan - This describes how the reserve can be maintained as a unit for the self-sustaining use of renewable natural resources by its residents. It lists non-predatory behaviour which is part of the culture of traditional communities and in conformity with Brazilian environmental legislation.

The plan provides a reference framework for checking that the use of resources is self-sustaining, does not harm the environment and respects current environmental legislation. It also helps IBAMA to check that environmental rules are properly complied with.

The plan explains which body will be responsible for its implementation and how this can be monitored.

Human activities in the environment are clearly defined and divided into categories: extractivist and farming (traditional activities), new activities (activities to be introduced, extracting new products in a model plot, such as fruit, oils, essences and even woods), activities with fauna, and activities in areas of common use.

The extractive activities section include such items as what types of extraction should be exploited, the ideal aera for each family to exploit, forms of exploitation, individual responsibilities in the area, precautions and care to be taken in the reserve, spatial division between extractivism and farming activities, size of the area for farming activities, use of cleared lands and pasture.

Main problems

Community problems

The community living in the Rio Ouro Preto extractive reserve usually has little schooling; its food and hygiene habits need improving, while its community organization needs to be made more functional.

Association problems

This too needs to be made more functional in the areas of planning and coordination, preparation, implementation and evaluation of projects, fund raising and financial administration, administration of community stores/canteens, upkeep of equipment, boats and engines; the association is very dependent on outside financial resources, and is little aware of the need to generate income to maintain itself in the future.

Current projects

- Training of assistant inspectors;
- Training of health agents;
- Recovery of popular medicine;
- Forest agriculture systems. Jointly with INPA, EMBRAPA and IEA, surveys are being made of current agro-ecological systems, with a view eventually to providing technical assistance and sponsorship for the introduction of household fruit and vegetable gardens and the introduction of demonstration units and observation units. This work involves seven communities, which have been given hundreds of cuttings and seeds. The aim is to strengthen the subsistence base of reserve residents, to lessen their dependence on goods of basic necessity traded in Guajará-

Mirim, to set up additional alternative sources of income by selling excess produce and to adopt soil conservation practices.

— Introduction of orchards. Two mixed orchards have already been planted, containing 23 species of fruit trees, including *pupunha, acerola,* mango, *araçá-boi,* coconut, *açaí, cupaçú* avocado and cashew nuts;

— Two cultivation units of leguminous manioc, in order to try natural weed control, green fertilizing and increased productivity;

— One unit with three varieties of sweet potato, the best variety of which will be selected and distributed;

— Distribution of vegetable seeds and technical assistance for the cultivation of egg plant, tomato, pepper, pumpkin, coriander, onion and greens;

— Introduction of two rice crops, Guaporé and BR 5102, with a view to carrying out productivity tests in comparison with currently used varieties.

Other socio-economic data

Production, prices and average annual gross income				
Rio Ouro Preto Extractive Reserve	Production	Price	Gross income	
	kg	US$	US$	Participation %
Agriculture			21,787.50	26.43
Rubber	70,000	0.60	42,000.00	50.95
Domestic animals	23,100	0.49	11,319.00	13.73
Hunting and fishing	15,925	0.46	7,325.50	8.89
TOTAL			82,432.00	100

Source: World Bank - IBAMA/CNPT - ECOTEC: RESEX project preinvestment study, 1994

Production, price and average gross annual income per family				
Rio Ouro Preto Extractive Reserve	Production	Price	Gross income	
	kg	US$	US$	Participation %
Agriculture			249.00	26.43
Rubber	800	0.60	480.00	50.96
Domestic animals	264	0.49	129.00	13.69
Hunting and fishing	182	0.46	84.00	8.92
TOTAL			942.00	100

Source: World Bank - IBAMA/CNPT - ECOTEC: RESEX project preinvestment study, 1994.

Some 750 persons in 123 families live in the Rio Ouro Preto Extractive Reserve and depend economically on the use of local natural resources *(photograph: Jeanine Levistre)*.

EXTRACTIVE RESERVES OF CIRIACO, MATA GRANDE AND EXTREME NORTH OF TOCANTINS

Miguel von Behr

Introduction

Chiefly because they are situated close to one another, within a few dozen kilometres, these three reserves are located in the same micro-region. This means they share the same geographic and environmental characteristics, and underwent the same establishment procedure. For the same reason they also share a similar socio-economic situation and have reached the same stage, even as regards establishment, legalization of land tenure and the social organization of communities. This means that their problems are also similar. In this sense, considering all these aspects, the comments below may be said to apply to all three reserves and at the end of each heading only the characteristics specific to individual reserves are mentioned separately.

Geographic and environmental data

In Brazil, the term *"babaçu"* is used to designate a series of palm species of the Orbignya genus, which occurs in continuous areas, from the state of Amazonas to southern Mato Grosso, and from Ceará to the state of Rondônia. Although it is not easy to estimate the exact area in which *babaçu* occurs, owing to its nature as an invasive species, in the states with the greatest concentration of *babaçu,* namely Maranhõ, Piauí, Mato Grosso, Tocantins and Goiás, according to a survey carried out by the Ministry of Industry and Commerce the palm covers a total area of some 185 000 km^2, with an effective coverage of 40%. According to the Mapping of occurrence and prospection of current *babaçu* potential in the state of Maranhõ (1981), this state has the greatest *babaçu* coverage area in Brazil, totalling 46 248.46 km^2.

According to AB'SABER (1971) and EITEN (1977), the *coco or babaçu* palm is a buffer vegetation in the transition from more open land (*cerrado*) to sparse forest (*caatinga*) to Amazonian forest, without belonging to any of these types. Some researchers have designated *babaçu* as a different type of vegetation (Feitosa, 1983; Ferri, 1974). Most of the current works on *babaçu,* however, in its present state of expansion consider it a form of secondary vegetation.

In decreasing order of area occupied, *babaçu* formations are distributed in sub-humid, tropical humid and tropical dry sub-humid climates. The most expressive example of the spread of *babaçu* in Maranhão is concentrated in the tropical-humid climate strip. The region where the reserves are located has a tropical, sub-humid climate. It is part of the Pré-Amazônia Maranhense. The *babaçu* vegetation cover (Orbignya martiana and Orbignya oleifera) occurs in the reserves either continuously or in groups, which may be either dense or spread out, along the waterways, exposed to flooding in the rainy period, or in depressions between the elevations, according to the physiognomic and geomorphological aspects of the regions in which they grow. *Babaçu* trees grow best on land where water is available in the soil throughout the year. The are also associated with the cleared lands and traditional agriculture, as practised by the extractive workers in artificial pastures. In the areas of former cleared forest (*capoeira*) and in abandoned pastures, there are large concentrations of *pindoveira* (young *babaçu*), which dominate other species. It is worth mentioning that the association of the *babaçu* trees and the *capoeira* cleared forest land is the one most frequently found in the state of Maranhão. The second type of formation, in terms of coverage area, is *babaçu* and pastureland, according to the Mapping of occurrences and prospection of the current *babaçu* potential in the state of Maranhão (1981).

The *babaçu* groves show characteristics of vegetation of anthropic origin, since it is considered to be pioneer or invasive forest, which means it colonises spaces after the forest has been degraded. They occur over large continuous areas, associated with various types of vegetation which succeeded pre-existing climax formations. According to the vegetation map of the environmental diagnosis of the state of Maranhão, prepared by the state's Secretariat for the Environment and Tourism (1981), the reserves are located in a region where the plant typology appears as a transition area between open non-deciduous seasonal forest and dense forest with *babaçu.*

The *babaçu* palm, which is present as one of the species in original native forests, has the ability to become the dominant species when the forest has been destroyed in order to expand the agricultural frontier and later turned into pasture.

The secondary forests, mainly the palms, therefore play an important role in the economic strategies of the extractive farmers, which are becoming increasingly vital, since the survival of these populations is becoming less and less viable in many developing countries.

On the other hand, the reserves, especially the Mata Grande reserve, still retain in smaller proportions areas of semi-forest (*cerrado*) and primary Amazonian forest, with other palms of economic value to the extractivists, such as *macúba (Acrocomia aculeata), buriti (Maurita flexuosa)* and *bacaba (Oenocarpos bacaba).*

In general, *babaçu* grows in soils of well drained, sandy, even clayey texture, over *latossolos,* indiscriminate concretions, hydromorphic aluminous and quartzy sand.

Ciriaco Extractive Reserve

This was set up by Decree No. 534 of 20 May 1992. It covers an area of 7 050 hectares and is located to the north of the municipality of Imperatriz, state of Maranhão, incorporating areas of the villages of São Francisco, Ciriaco and Alto do Peba in the western part of the state.

The area may be accessed by land in two ways: either by the road linking the village of São Domingos (located on the main Imperatriz-Cidelândia highway) to the urban centre of Viração (located on the banks of the Tocantins river, south of Ciriaco), or by the Belém-Brasilia highway, by driving up to the 50 km mark, from Imperatriz, and then going on to Cidelândia and São Domingo.

An alternative access, although the reserve is entirely on dry land, is by the Tocantins river, starting from the locality of Viração.

Mata Grande Extractive Reserve

Mata Grande was set up by Decree No. 532 of 20 May 1992. It is located in the municipality of João Lisboa and Imperatriz, in the state of Maranhão, incorporating the villages of Mata Grande and Água Viva in the south west of the Arapari mountains, in the western part of the state of Maranhão.

The reserve, which covers 10 500 hectares, may be accessed in three ways: either by the road linking Água Viva to Vila Dimas (a neighbourhood of Imperatriz); by Vila Bananal, situated at the 30 km mark on the Belém-Brasilia highway, driving to the village of Mata Grande, or by the road running between Imperatriz, João Lisboa and Mucuiba, also through Mata Grande.

Considering the poor maintenance of roads in the region, the best access is through João Lisboa, even though it is the longest way around.

Apart from stretches of the Belém-Brasilia highway and the road from Imperatriz to João Lisboa, access roads to the area are unstable and are barely suitable for traffic during the rainy "winter" period.

Extractive reserve of the extreme north of Tocantins

Decree No. 535 of 20 May 1992 provides the legal basis for this reserve, which covers 9 280 hectares. It is situated to the north of highway TO-496 and to the south of the Tocantins river, covering practically all the municipality of Carrasco Bonito, at the extreme north of the state of Tocantins.

The best access is through the town of Augustinópolis bearing in mind that, as in the case of the other two reserves, the roads entering the reserve are not easy to drive on, especially during the rainy period.

History of the reserves

The extractivists in the region have organized in a number of Rural Workers' Unions to defend their interests.

The idea of setting up the extractive reserves arose from the need to preserve the palm tree as a genetic heritage, and a rich and natural source of vegetable oil, as well as in response to the demands of the extractive workers for access to natural resources, on which they survive. This then led to the initiation of appropriate studies in preparation for setting up the reserves.

Apart from an aerial survey of the region, field visits were carried out for the purpose of identifying villages next to and inside the proposed areas. On the basis of these surveys, the area with the greatest continuous density of *babaçu* was selected.

Based on the assumption that it was the oldest people in the region who had accumulated most knowledge, on arriving in each village the researchers sought out the oldest resident and, after the usual introductions, began to interview him, in order to obtain more information about the region, finding out how it had been occupied and what changes had occurred in the ecosystems.

The suitability of areas for setting up an extractive reserve was also discussed, amongst other matters, in meetings with leaders of workers' unions and with the actual communities in the region. Several areas were finally selected for further analysis.

Ciriaco Extractive Reserve

The following governmental and non-governmental organizations participated in and supported the establishment of the Ciriaco extractive reserve:

- General Procurator's Office;
- State Secretariat for the Environment and Tourism of Maranhão;
- Institute of Amazonian and Environmental Studies;
- Imperatriz Rural Workers' Union;
- Rural Workers' Education and Culture Centre (CENTRU);
- Santa Tereza D'ávila parish;
- Caritas Brasileira.

Mata Grande Extractive Reserve

The following governmental and non-governmental organizations participated in and supported the establishment of the Mata Grande extractive reserve:

- General Procurator's Office;
- State Secretariat for the Environment and Tourism of Maranhão;
- Institute of Amazonian and Environmental Studies;
- Imperatriz Rural Workers' Union;
- Rural Workers' Education and Culture Centre (CENTRU);
- Santa Tereza D'ávila parish;
- Caritas Brasileira.

Extractive reserve of the extreme north of Tocantins

The following governmental and non-governmental organizations participated in and supported the establishment of the extractive reserve of the extreme north of Tocantins:

- General Procurator's Office;
- Fundação Natureza do Tocantins (Nature Foundation)- NATURANTINS;
- Comissão Pastoral da Terra (Pastoral Land Commission)- CPT;
- Institute of Amazonian and Environmental Studies;
- Axixa do Tocantins Rural Workers' Union;
- Federação dos Trabalhadores na Agricultura do Estado do Tocantins (Agricultural Workers' Federation)- FETAET;
- Augustin-ópolis Workers' Union;
- Fundação de Ciência, Aplicação e Tecnologia Espaciais (Space Science, Application and Technology Foundation) - FUNCATE;
- Itaguatins Rural Workers' Union;
- Praia Norte Rural Workers' Union;
- Araguatins, Buriti, São Sebastião and Esperantina Rural Workers' Union.

Socio-economic data

According to the preliminary socio-economic survey commissioned by the CNPT in 1992, among the extractive resources on which the viability of the extractive reserves depends, the *babaçu* palm ranks on an equal footing with the Amazonian rubber tree, owing to its widespread geographic distribution in the pre-Amazonia and north-east areas of Brazil and owing to the fact that it sustains an economic activity - the processing or breaking of the *babaçu* coconut - on which some 450 000 families depend (IBAMA, 1990). In other words, the socio-economic importance of *babaçu* may be measured by the number of families which depend on it directly. The decline of the extractive *babaçu* industry has been due to a number of reasons, including the low price of the kernel, the systematic transformation of *babaçu*-growing areas into pastureland, inefficient processing techniques and over-scaled projects. This decline has given rise to some terrible consequences not only for the industries involved, but also for the rural workers, who depend on *babaçu* for their survival. Considering that less than 20% of total *babaçu* production is marketed at present, the extraction of the species offers an enormous potential as a sustainable alternative to deforestation (Anderson, 1992). Although the greatest technological obstacles which existed as regards the processing of *babaçu* have now been overcome, the failures of the past suggest that the most appropriate solution lies in small-scale industries, where investment and operating costs are minimal, apart from the fact that they are more suitable from a social point of view. According to the minutes of the Workshop on political, social and technological alternatives for the sustainable development of *babaçu,* which was held in 1992, the social significance of *babaçu* is even greater because the product is exploited in the interval between the harvesting of the main crops, so that it provides a means of maintaining local families and restraining the rural exodus.

Babaçu gives Brazil the greatest oleaginous potential in the world. The babaçu industry was already at one time the leading world industry based on the processing of a single native plant product. In addition, the *babaçu* plant is held to be one of the most efficient producers of biomass among forest species in sub-humid tropical ecosystems (Anderson,

1983). It is estimated, however, that 90% of the native production of nuts from these palm trees remains unused in the forest each year. This wastage is due to the small producer's lack of access to the land, in other words, to the property rights exercised by large landowners over natural resources, which determine the respective shares of the benefits obtained from exploiting the nuts. The agrarian and agricultural structure which has altered and hampered the small producers' access to the *babaçu* groves has also affected the supply of raw material to the oil industry, although the impact on the small producer's survival strategy is even more direct.

Nowadays, with increasing development in the region, combined with the prevalence of latifundiarism and large-scale cattle raising, a large proportion of the population, descended from indigenous peoples, negroes and migrants from the semi-arid north-east, is compressed into extremely poor villages, situated within the only physical space left to them in that huge region, namely the narrow band of land stretching between the highways and the outer fences of the large farms. The houses are made with mud walls (*taipa*) and are covered with palm leaves, with practically nothing inside. It is on these strips of bare, unproductive land, very often without any water, that a huge population of workers resists submission in the hope of seeing their right to live one day guaranteed by the state.

At the heart of this resistance lies the *babaçu* nut, which is gathered on the farms by the men and broken up in the homes by the women, who are known as the *quebradeiras de coco* (coconut breakers). Caught in a medieval type of system, the development of this productive activity can only be understood and explained in terms of the pressing and inescapable need which arises each day in the struggle for survival.

Fenced in by the farmers, who exploit their labour on poorly paid, seasonal tasks, such as clearing forests and pasture land, the majority of the people, apart from giving up their rights of possession to speculators, only manage to plant a few subsistence crops on cleared land (rice, maize, *feijão* bean and manioc), against payment of an exorbitant rent to the landowner (up to 50% of output value), also on condition that, after harvesting, the area is turned into pastureland. According to the land register and socio-economic survey of the reserves carried out in 1993, about 80% of the heads of family have land clearance as their main economic activity and in terms of occupation almost half the people using the reserves are tenants. A significant number of people, in other words, do not even benefit from this "handout" by the landowners, and simply have no means of producing for their subsistence. Moreover, as the pasturelands expand, the places where the *babaçu* nuts are gathered are gradually becoming increasingly distant.

The coconut, then, is practically the only alternative left to many families and even to whole communities and it is the most important productive activity for everyone living in the *babaçu*-growing areas. According to the land register and socio-economic survey of the extractive reserves mentioned earlier, over half the women gather *babaçu*, especially in the reserve of the extreme north of Tocantins (80.6%), while practically 90% of those interviewed in the three reserves quote *babaçu* as their main extractive product. The violence to which the workers and their families are exposed to obtain the nuts on which they depend for survival is therefore disturbing. The women who break up 10 kg of *babaçu* exchange it at the end of the day for less than 2 kg of rice (Assema, 1990). Furthermore, in many areas the coconut breaking is carried on in a practically clandestine fashion, because many of the farmers do not allow people on their land to gather the nuts.

The breaking of the *babaçu* coconut and subsistence agriculture therefore constitute the main economic activities in the area. The sale of family labour to the farmer is one of the few ways of obtaining any income in the region, which the worker has to accept out of absolute necessity and for lack of any alternative.

The region mainly produces nut kernels, rice, maize, *feijão* beans and manioc, offering an income below the minimum wage (less than US$ 80.00), which obliges the local communities to submit to the system established by local farmers. Working as day labourers, while it is a means of improving the sparse family income, leaves the workers tied to the latifundia system. According to the preliminary socio-economic survey carried out in 1992 at the CNPT's request in preparation for the establishment of the reserves, it is estimated that, if the workers had free access to the lands they work, their present income, under the same prevailing conditions of production and marketing, would be three times greater than at present.

Despite its great energetic and nutritional potential, apart from the kernel itself, practically all the rest of the fruit (the epicarp, mesocarp and endocarp) obtained from manual breaking is wasted, in other words, almost 90%. The following uses could, however, be exploited: direct consumption, extraction of milk, edible oil, production of soap, cosmetics and lighting, sludge used as animal feed, insect larvae extracted from the kernels and used for human consumption, bait, shells, which produce vegetable charcoal of excellent quality, smoke against insects, tar, handicrafts, textile fibres, alcohol, wall and floor coverings, etc. According to the figures, for 240 000 tons of kernels, the annual wastage in other respects amounts to 800 000 tons of amylaceous mesocarp, 2.9 million tons of fibrous epicarp and ligneous endocarp, equivalent to 1 million tons of coal with 85% fixed carbon, 280 000 m^3 of anhydrous alcohol, 234 000 tons of methanol and 184 000 tons of basic ration and other products (IPES, 1983, in Diagnosis of main environmental problems of the state of Maranhão, 1991).

Meanwhile, the import of oil from Malaysia by major purchasers, such as Gessy Lever and Colgate Palmolive, has upset the equilibrium of regional *babaçu* oil manufacture. As a result, the oil imports have come to act as a price regulator in the market for vegetable oils (Assema, 1993). Compared with *babaçu* oil, Malaysian oil arrives at Brazilian ports at practically half the price, including all duties. With the reduction in tariffs, industries are now having problems negotiating their products on the domestic market, according to the diagnosis carried out by the Workshop on *babaçu* held in São Luis in 1992. The small agro-extractive cooperatives are not in a position to build up a sufficient volume of production and capital, since they are exposed to the variations of local oil producers. Besides, owing to the uncertainty of the dry season in some years, the periods of harvesting and between harvests have tended to vary, even within micro-regions, with the differences being more marked among areas with a greater proportion of cleared land or degraded land (abandoned pastures).

According to the Mapping of occurrences and prospection of current *babaçu* potential in the state of Maranhão (1991), productivity in the state averages around 1 804 kg/hectare, with a 48.8% average percentage of productive palm trees per hectare. There are some exceptions, ranging from areas with no nut production at all to some with 10 500 kg/ha/year. According to agronomist and EMBRAPA/EMAPA researcher José Mário Ferro Frazão, although the palm has never been domesticated and takes 15 to 20 years to reach maturity, its annual productive potential is 13 million tons of coconut. According to the Diagnosis of the main environmental problems of the state of Maranhão (1991), the state has an estimated area of *babaçu* occurrence as high as 10 300.5 palms per 1 000 hectares, which amounts to a 54.2% share of all *babaçu* plantations in the country. According to the same diagnosis, Maranhão's average output per hectare can vary between one and three tons, with peaks of up to seven, 10 or even 15 tons per hectare per year. In order to obtain 10 kg of nut kernels, for instance, 120 kg of whole coconuts are used, while on average a skilful breaker can extract up to 15 kg in 8 hours of work, the state's average being around 5 kg per working day (Assema, 1990).

The low volume marketed within the area of influence of the three reserves is due in part to the lack of efficient organization on the part of the communities when it comes to marketing the produce. Despite some experience acquired with community stores, marketing as such is left to the *"marreteiros"* (middlemen), who generally, even when they live in the villages, are not workers like the others. They are part of a marketing chain and often live in nearby towns, only rarely putting in an appearance in the communities. This intermediate marketing stage ends when the products are finally offered on the major regional markets (Imperatriz and Açailândia). The extractive workers also take whatever products they can to the roadside to sell, in order to obtain some ready cash to buy medicines and basic necessities.

The village families buy their own supplies in a few *"bodegas"* (small stores), which generally belong to the middlemen, and where the local residents can find basic necessities (such as sugar, salt, coffee, etc.) for very high prices, which keeps the workers chronically in debt to the *marreteiros.* As the communities become better organized, an alternative to this dependence on the middlemen has appeared within the reserve area, in the form of the introduction of community stores (*cantinas*) kept by associations of women nut breakers. In the community stores, basic products can be exchanged for *babaçu* on better price terms for the extractive workers. The nuts are then sold on by the local associations to purchase more products, such as sugar, matches, salt, kerosene, etc. In a number of places, such as Lago do Junco and Imperatriz, in Maranhão and in Sítio Novo and São Miguel, where the Sete Barracas agricultural club is situated in the state of Tocantins, new agro-extractive cooperatives linked to Assema unions and other basic groupings are trying to set up decentralized ways of processing and marketing *babaçu.* These initiatives, which are supported by the CNPT, offer the possibility of diminishing transport costs and increasing the earnings of *babaçu*-dependent residents.

It may be added that, when *babaçu* prices begin to rise, thanks to this work of the community stores, big farmers close to the villages which have the most productive stores begin to recruit coconut "pickers" and send the nuts to be broken in the towns, for half the price. On a number of occasions, groups of nut-breakers in the three reserves have launched "stalemate" operations, closing the roads and stopping the nut trucks from leaving the area. According to the extractive workers, they are not opposed to the nut breakers in town having access to the babaçu, but tagainst the farmers who submit workers to humiliating treatment. Similarly, within the general reserve area, a number of *"empates"* (stalemate operations) have already been conducted against *babaçu* devastation, similar to those in Amazonia, organized by the nut breakers.

The view defended by the agro-extractive workers is that, while they recognize that the nuts are legally speaking on the property of the farmers, they are really fighting for FREE *BABAÇU:* nobody planted it; nobody grew it; it is wealth which belongs to the nation, like the rivers, like the air, beyond the rules of private property. In the workers' view, the social function of property, which is guaranteed by the Federal Constitution, does not apply only to the land, but to other natural resources as well.

Apart from the products extracted from the *babaçu* palm, the workers exploit other native plants in the reserve, such as *pati* (the kernels of which can be mixed with those of the *babaçu)* and *inajá* (a coconut which is eaten).

The main medicinal species of the forest used by local people are the: *imbaúba* (Cecropia palmata), *tipi* (a type of jute), *mastruz* (cress), *jambu* (Spilanthes oleracea), *jaborandi* (Pilocarpus jaborandi) and *canapu* (Physalis heterophylla).

Hunting, which was always important in the area, continues to be one of the main sources of animal protein for the population. Species which still occur include the *caititu* (peccary), deer, capybara, *tatu* (armadillo), *cutia* (agouti), tapir, etc.

Fishing is also important for the local population, the most consumed species being *surubim, curimbata, piau, matrinchã and traíara.*

Most of the villages within the area of influence of the reserve have a school with teaching only up to the first grade, operating under precarious conditions. Facilities are very poor and the teachers, mostly laymen, are poorly paid.

Despite the fact that most of the villages have dispensaries, the service provided is completely inadequate, in terms both of medicines and of doctors. Some villages have health agents, but there is no public transport to remove patients.

Most of the villages are connected to the electricity grid, although the current is still not accessible to most of the resident families, owing to their extremely low purchasing power.

Very few villages have any public transport system.

Ciriaco Extractive Reserve

Almost 200 families are linked to the reserve and the villages most directly dependent on it are Ciriaco, São Francisco and Alto do Peba.

According to the map and survey of the potential for *babaçu* occurrences (1982), the extractive reserves of Ciriaco and Mata Grande are located in a *babaçu* patch of 241 500 hectares with 70% coverage (169 050 hectares), with an average concentration of 66 adult palms per hectare and 46% productive palms. The average is 31 productive palms per hectare. The adult palm concentration refers to the average number of adult palms detected per module in the patch. Average productivity is 1 364.1 kg/ha and *babaçu* production per patch, where the Ciriaco and Mata Grande reserves are situated, is around 230 601.1 tons.

Mata Grande Extractive Reserve

Approximately 150 families depend on the reserve's natural resources and live in the villages of Mata Grande, Água Viva and São Luiz.

Extractive reserve of the extreme north of Tocantins

Around 60 families use the reserve for their survival, living in the following villages: Centro do Gonçalo, Centro Firmino, Vinte Mil, Cachimbo and the municipal centre of Carrasco Bonito.

According to the same source quoted for the Ciriaco reserve, the reserve of the extreme north of the state of Tocantins lies within a *babaçu* patch of 222 250 hectares, with 40% coverage (88 900 hectares), with an average adult palm concentration per hectare of 54 species and 12% productive palms. As in the case of the other two reserves, the adult palm concentration refers to the average number of adult palms identified per module in the patch. Average productivity is 751.8 kg/ha and *babaçu* production per patch where the reserve is located is around 66 835 tons.

Social organization

As had already been mentioned in the socio-economic survey drawn up prior to the establishment of these reserves, social organization within their areas of influence is still insufficient to ensure the sustained management of reserve resources.

In the villages, the job of setting up a specific association for the reserve, despite some advances and a great deal of effort spent by local leaders, has run into difficulties because people tend to confuse associations, unions and collective forceful land occupation. Besides, according to local people, the fact that IBAMA has not expropriated the land makes it difficult for the residents to organize at all, since there is considerable discrimination on the part of the owners towards workers who try to organize; this takes the form of preventing them from planting or extracting *babaçu* from their lands, apart from other reprisals. A further obstacle to the efforts of small *babaçu* producers to include the usufruct over the palm trees among their rights has been the fact that there are a great many people looking for work in the region. As a result, many of the workers tend to accept the conditions of work imposed by the bosses, to the detriment of those who are a little better organized and are trying to obtain greater respect for their rights (Contag, 1983, in May 1990).

The illegal possession of land *(grilagem)* and cattle raising are very gradually, step by step, winning the land from babaçu extraction, giving rise to constant social conflicts.

In every village of the three reserves, there is a union delegate who is responsible for coordinating and mobilizing the workers, since the unions are the main defenders of their interests.

According to the land register and socio-economic survey of the reserves (1993), the best organized agro-extractive workers are those in the reserve of the extreme north of Tocantins, where 83% of those interviewed belonged to some union or association. In the other two reserves, only about half the people interviewed belonged to one.

Ciriaco and Mata Grande Extractive Reserves

The institutions working in the communities of these two reserves for organization purposes are: the Imperatriz Agro-Extractive Workers' Cooperative, Imperatriz Rural Workers' Union, Association of Women *Babaçu* Coconut Breakers, Rural Worker's Education and Culture Centre (CENTRU) and the Comissão Pastoral da Terra, all with offices or representatives in Imperatriz, which is situated a few kilometres away from the reserves.

Extractive reserve of the extreme north of Tocantins

The Augustinópolis Rural Workers' Union, the Sete Barracas Agricultural Club, the Association of Women *Babaçu* Coconut Breakers and the Comissão Pastoral da Terra are practically the only institutions working with the extractive workers in this reserve, in support of their organization.

Utilization Plan

The multiple use of natural resources currently practised by the agro-extractive workers in the region offers a production model which has the advantage of exerting less pressure on the land.

Extractive activities, hunting, fishing and small-scale subsistence agriculture were harmoniously integrated, before the environment suffered the impact of the activities of the large landowners.

The small producers make the best use of this forest ecosystem, because they are the most efficient in handling the *babaçu* palms, combining coconut breaking with other productive activities available in these areas, especially those related to food production.

In order to increase the productivity of *babaçu* trees, it is recommended that a selective thinning of unproductive palm trees should be operated in the areas, down to levels of 200 palm trees per hectare, which is considered to be a dense *babaçu* grove according to the 1981 mapping of occurrences and prospection of the current *babaçu* potential in the state of Maranhão. Another way of increasing *babaçu* productivity is to control the overpopulation of young plants and combining *babaçu* with other perennial plants adapted to the region.

In areas where the palm density is around 80 plants per hectare, which incidentally is the average occurrence in the three reserves, according to the 1981 mapping referred to earlier, an effort should be made to recover these degraded *babaçu* groves, again by thinning out excess pindovas in existing areas, trying to leave only about 30 of the more vigorous plants interspersed in existing empty spaces.

According to the conclusions of the aforementioned 1981 mapping, the variations in *babaçu* productivity are not necessarily correlated with density, since dense *babaçu* groves may have either high, medium or low productivity, which is also found to be the case in average density or thinly planted groves, in proportion.

Babaçu groves may also be recovered where trees have been cut down, by nursing the *pindova* population over a period of 15 to 20 years, the time which it needs for these to become productive adult *babaçu* palms.

The low productivity of some *babaçu* areas is due to the fact that almost 50% of the palm trees are practically unproductive, that is, produce only male inflorescence, and that they are exposed to "stress" caused by inadequate handling. In addition, the palm trees are subject to considerable competition from other associated plant species, and to the overpopulation of young plants.

Through methods of rational management involving simple practices, such as the felling of unproductive palm trees and *pindovas,* the *babaçu* groves in the three reserves could achieve a productivity increase of some 50% over three to four years, according to observations made in the socio-economic survey on which the establishment of the reserves was based.

Current projects

The land survey of the extractive reserves has now been completed with a view to settling the legal land problems related to expropriation. This work has made it possible to evaluate the cost of expropriating the land and land improvements, although the funds have not been deposited with the courts owing to a shortage of resources. Needless to say, the owners put up many obstacles and considerable resistance in their efforts to hamper the work of evaluation, which was firmly supported by the Instituto Nacional de Colonização e Reforma Agrária. Despite the fact that the decree establishing the reserves allows the entry of government officials in the reserves, in order to ensure the safety of its assessors IBAMA had to go to court to gain access to the properties before the evaluation work could be completed.

In the course of expropriation procedures undertaken for the three reserves, the time allowed for starting the necessary legal proceedings under current legislation elapsed. Nevertheless, the areas continue to be extractive reserves, since after one year, i.e. by May 1995, the decree may be reissued and the expropriation procedure reinitiated.

Main problems

1. Little community organization, caused by a poor understanding on the part of residents of the differences between unions, associations and forceful collective land occupation.
2. Constant social conflicts brought about by unlawful appropriation of properties *(grilagem).*
3. The social activities of communities tend to be intermingled with religious and party political activities.
4. Lack of any specific policy for products based on *babaçu* extraction, such as establishing lines of credit, adapted to the cultural, ecological, economic and social needs of extractive workers.

Buriti fruit *(Mauritia flexuosa) (photograph: Julio Ruiz Murrieta).*

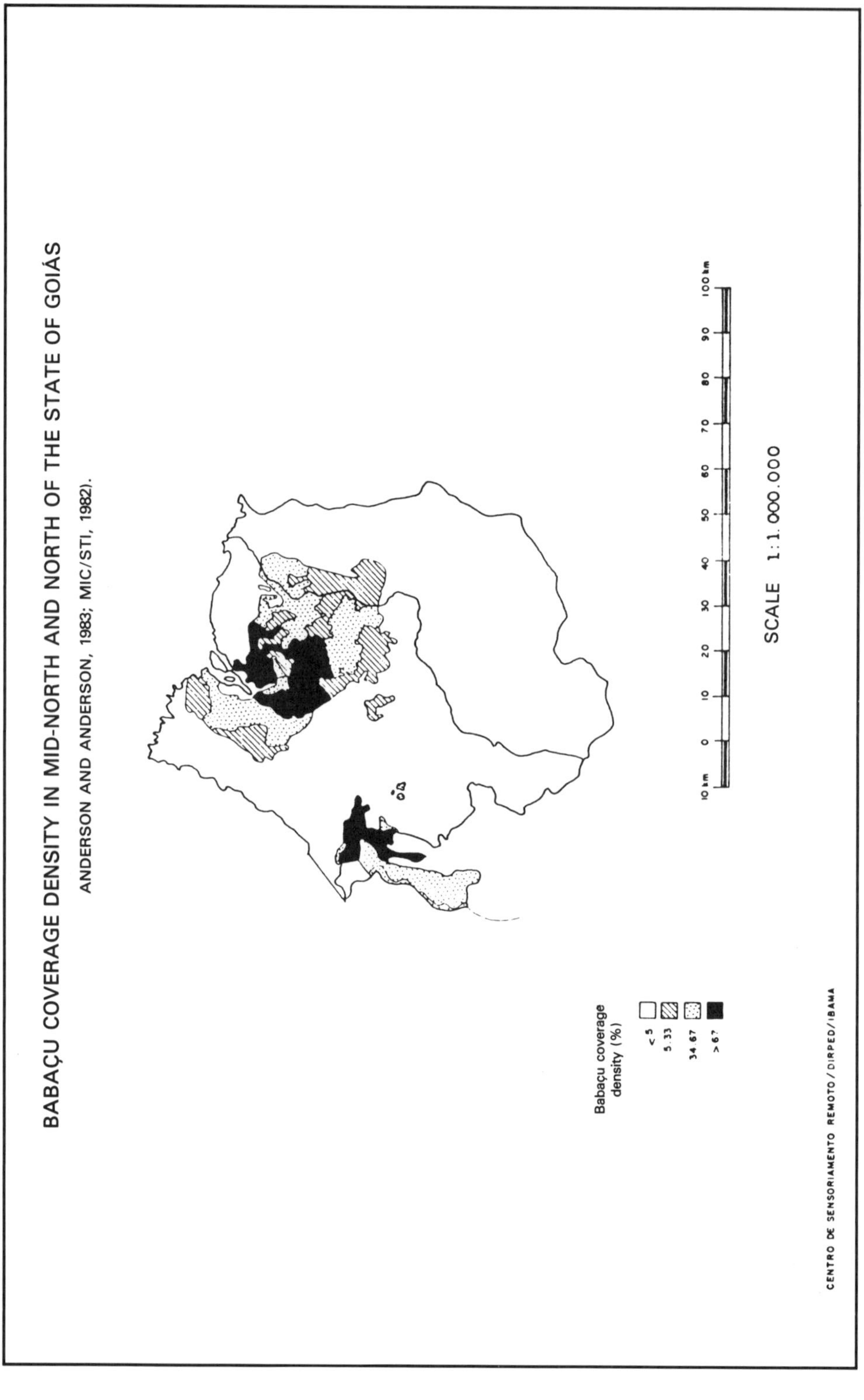
BABAÇU COVERAGE DENSITY IN MID-NORTH AND NORTH OF THE STATE OF GOIÁS
ANDERSON AND ANDERSON, 1983; MIC/STI, 1982).
Babaçu coverage density (%)
<5
5.33
34.67
>67
10 km
0
10
20
30
40
50
60
70
80
90
100 km
SCALE 1:1.000.000
CENTRO DE SENSORIAMENTO REMOTO/DIRPED/IBAMA

AREAS OF BABAÇU OCCURRENCE in GO, TO, MT, PI e MA

SOURCE: MAPPING AND SURVEY OF POTENTIAL OCCURRENCES OF BABAÇU IN THE STATES OF MARANHÃO, PIAUT, MATO GROSSO AND GOIÁS. MINISTRY OF INDUSTRY AND COMMERCE/SECRETARIAT FOR TECHNOLOGY AND INDUSTRY.

CENTRO DE SENSORIAMENTO REMOTO/DIRPED/IBAMA

LOCATION OF EXTRACTIVE RESERVES IN BRAZIL

CENTRO DE SENSORIAMENTO REMOTO/DIRPED/IBAMA

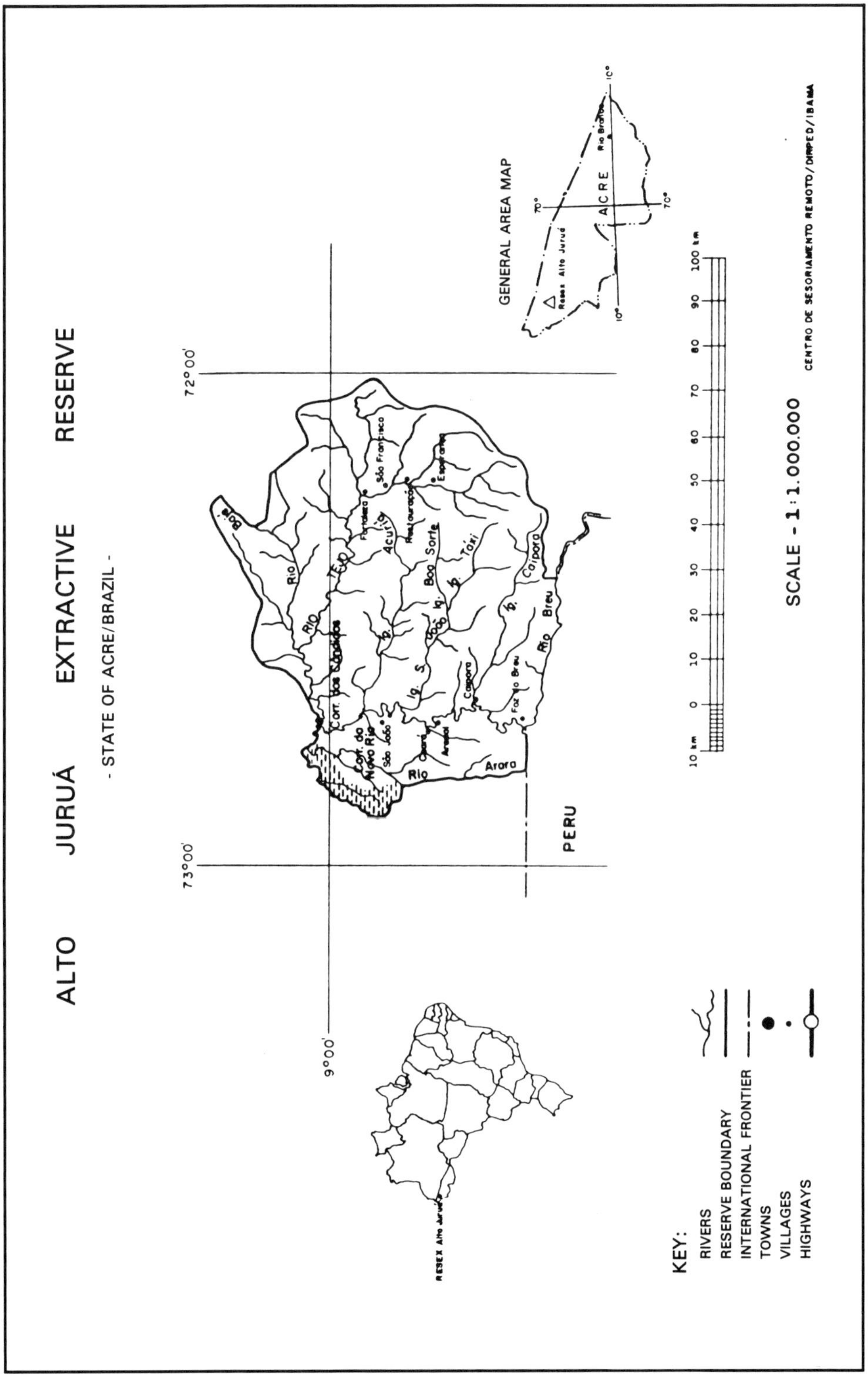
ALTO JURUÁ EXTRACTIVE RESERVE
- STATE OF ACRE/BRAZIL -
73°00'
72°00'
9°00'
PERU
GENERAL AREA MAP
ACRE
SCALE - 1:1.000.000
KEY:
RIVERS
RESERVE BOUNDARY
INTERNATIONAL FRONTIER
TOWNS
VILLAGES
HIGHWAYS
CENTRO DE SESORIAMENTO REMOTO/DIRPED/IBAMA

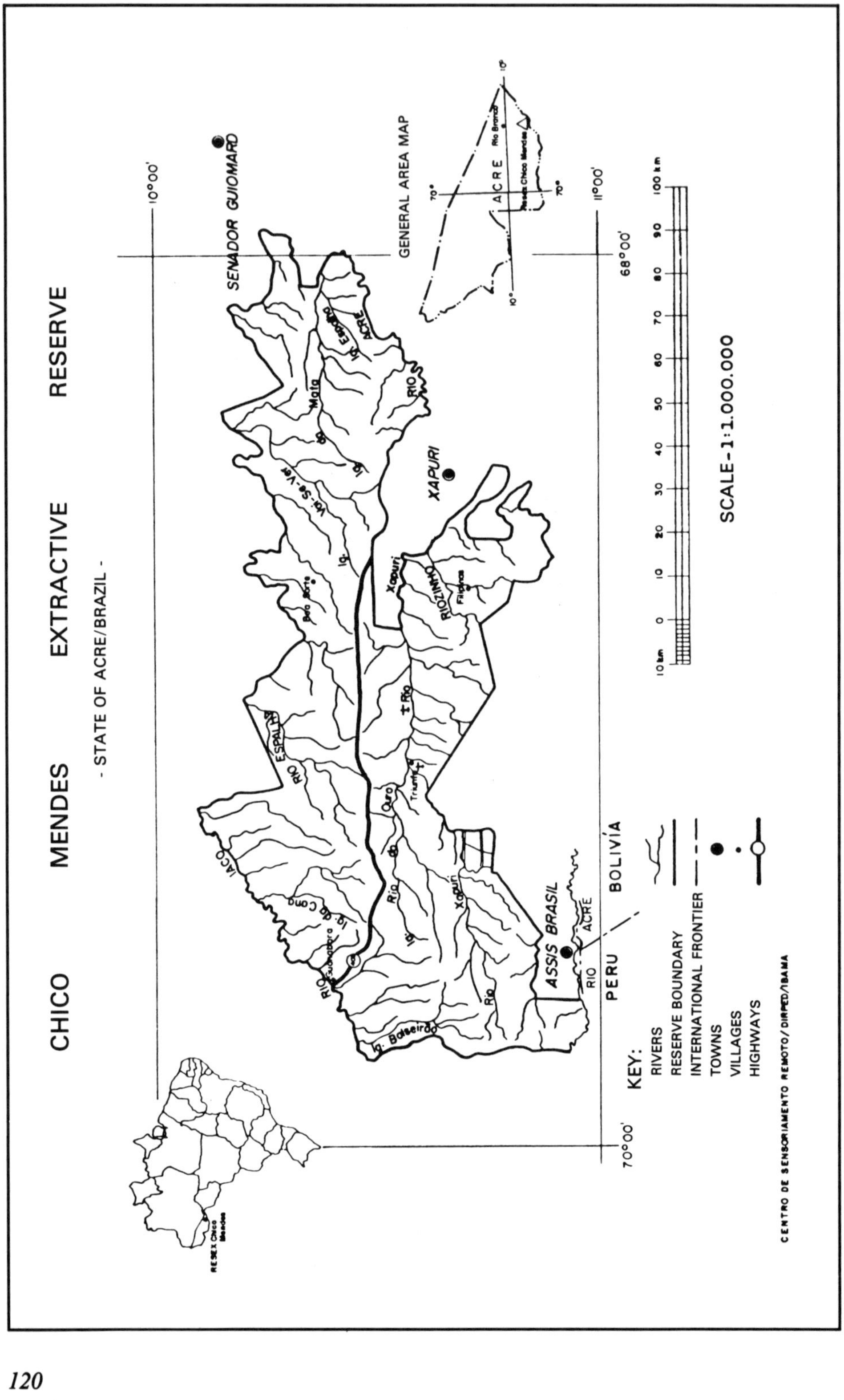
CHICO MENDES EXTRACTIVE RESERVE
- STATE OF ACRE/BRAZIL -
10°00'
11°00'
70°00'
68°00'
SENADOR GUIOMARD
XAPURI
ASSIS BRASIL
PERU
BOLIVÍA
RIO ACRE
RIO IACO
RIO ESPALHA
RIOZINHO
Xapuri
GENERAL AREA MAP
ACRE
Rio Branco
SCALE-1:1.000.000
KEY:
RIVERS
RESERVE BOUNDARY
INTERNATIONAL FRONTIER
TOWNS
VILLAGES
HIGHWAYS
CENTRO DE SENSORIAMENTO REMOTO/DIRPED/IBAMA

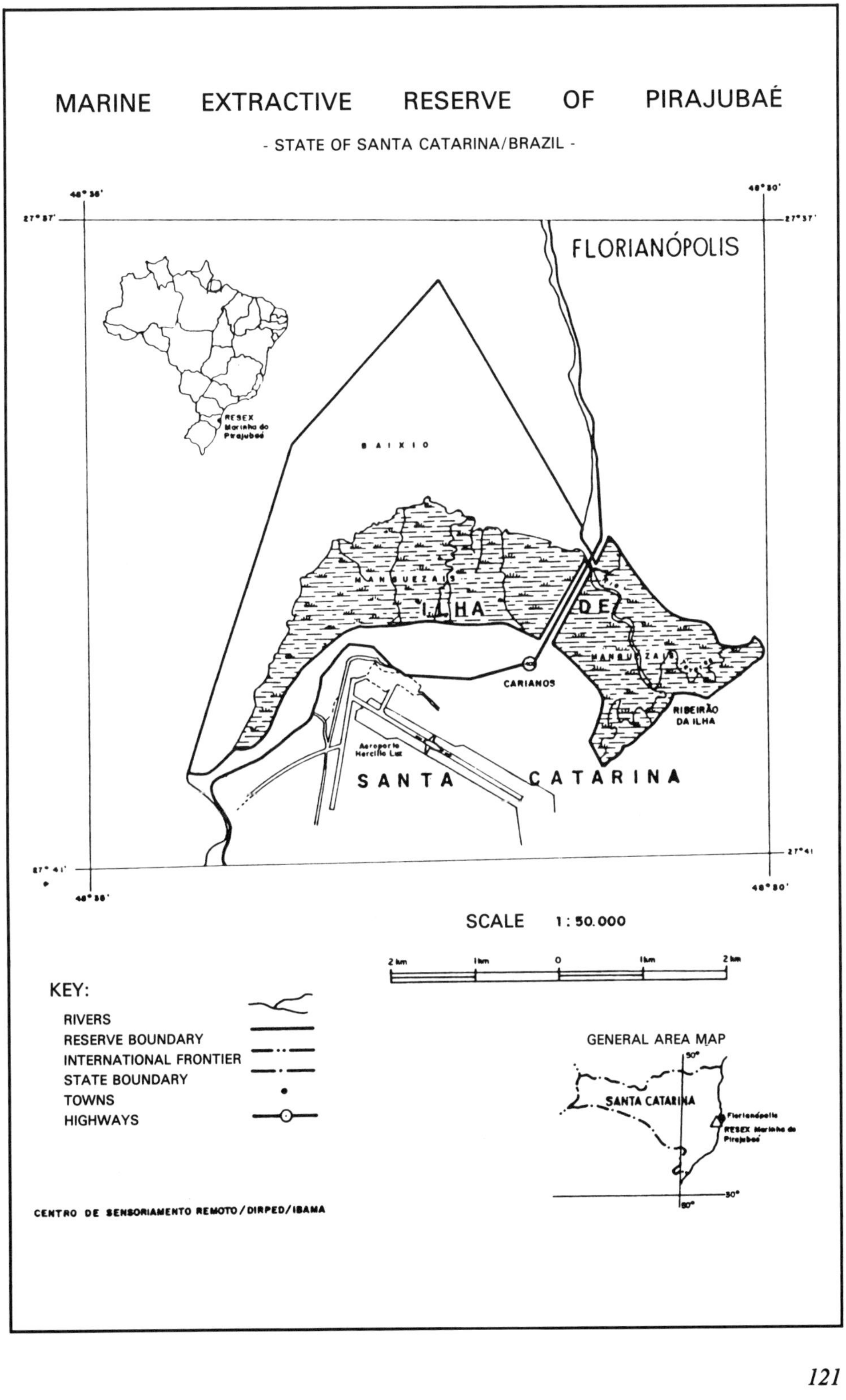
MARINE EXTRACTIVE RESERVE OF PIRAJUBAÉ
- STATE OF SANTA CATARINA/BRAZIL -
FLORIANÓPOLIS
BAIXIO
MANGUEZAIS
ILHA DE
CARIANOS
RIBEIRÃO DA ILHA
SANTA CATARINA
SCALE 1: 50.000
KEY:
RIVERS
RESERVE BOUNDARY
INTERNATIONAL FRONTIER
STATE BOUNDARY
TOWNS
HIGHWAYS
GENERAL AREA MAP
SANTA CATARINA
CENTRO DE SENSORIAMENTO REMOTO/DIRPED/IBAMA

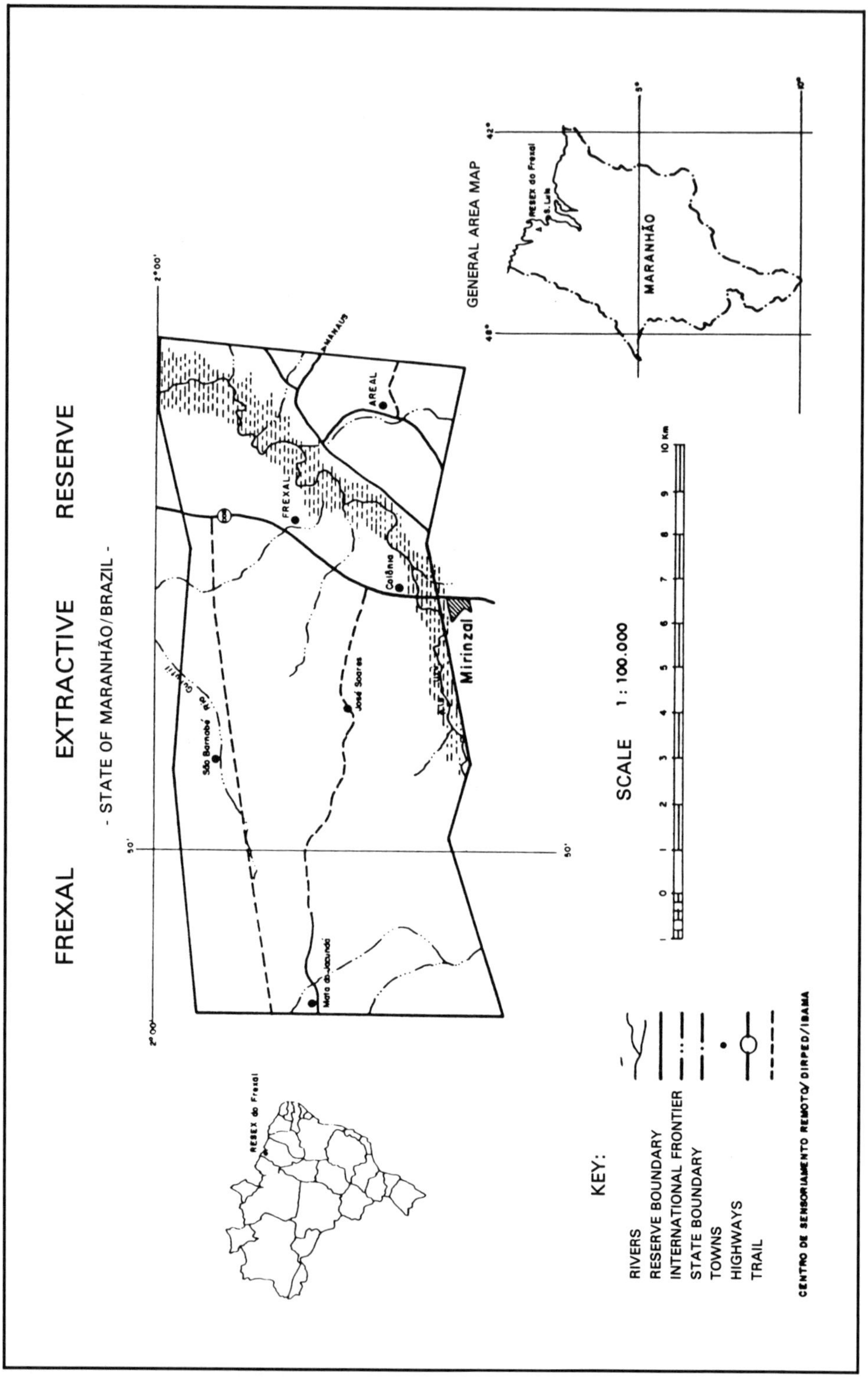
FREXAL EXTRACTIVE RESERVE
- STATE OF MARANHÃO/BRAZIL -
GENERAL AREA MAP
MARANHÃO
RESEX do Frexal
FREXAL
AREAL
Colônia
José Soares
São Barnabé
Mirinzal
SCALE 1 : 100.000
KEY:
RIVERS
RESERVE BOUNDARY
INTERNATIONAL FRONTIER
STATE BOUNDARY
TOWNS
HIGHWAYS
TRAIL
CENTRO DE SENSORIAMENTO REMOTO/ DIRPED/IBAMA

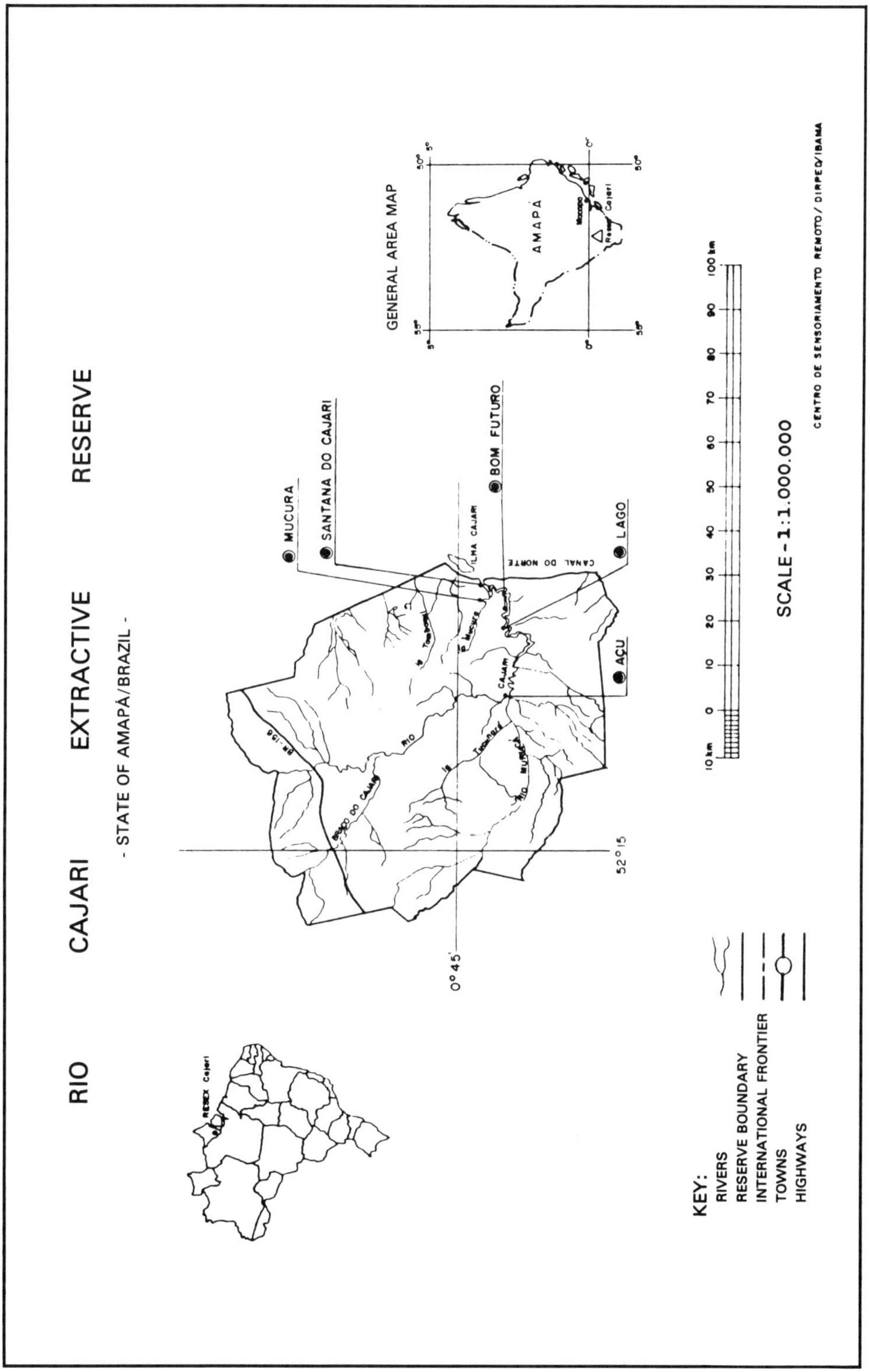
RIO CAJARI EXTRACTIVE RESERVE
- STATE OF AMAPÁ/BRAZIL -
MUCURA
SANTANA DO CAJARI
ILHA CAJARI
BOM FUTURO
CANAL DO NORTE
LAGO
AÇU
BR-156
0°45'
52°15
GENERAL AREA MAP
AMAPÁ
10 km 0 10 20 30 40 50 60 70 80 90 100 km
SCALE - 1:1.000.000
KEY:
RIVERS
RESERVE BOUNDARY
INTERNATIONAL FRONTIER
TOWNS
HIGHWAYS
CENTRO DE SENSORIAMENTO REMOTO / DIRPED/IBAMA

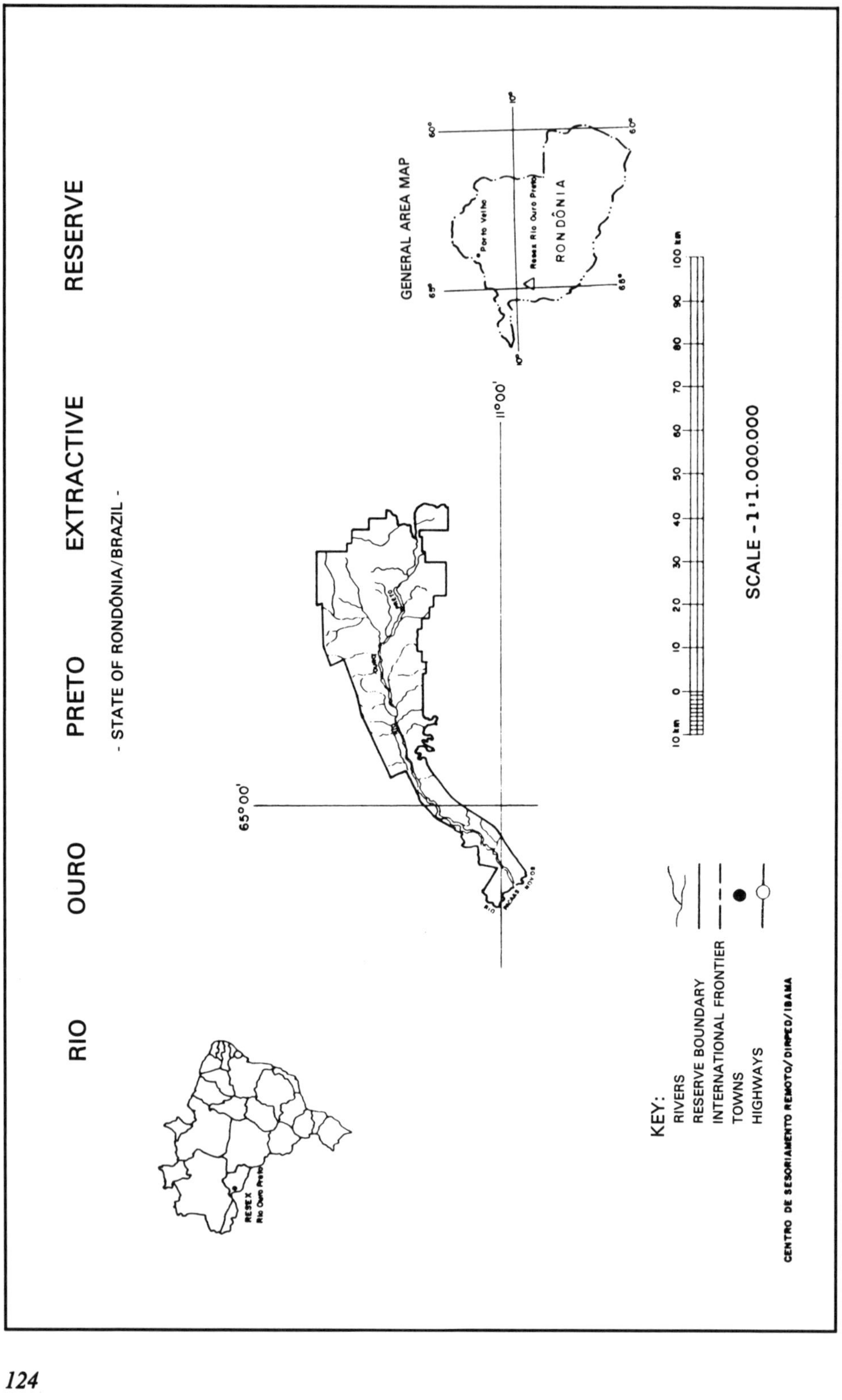
RIO OURO PRETO EXTRACTIVE RESERVE
- STATE OF RONDÔNIA/BRAZIL -
65°00'
11°00'
RESEX
Rio Ouro Preto
GENERAL AREA MAP
Porto Velho
Resex Rio Ouro Preto
RONDÔNIA
10 km 0 10 20 30 40 50 60 70 80 90 100 km
SCALE - 1:1.000.000
KEY:
RIVERS
RESERVE BOUNDARY
INTERNATIONAL FRONTIER
TOWNS
HIGHWAYS
CENTRO DE SESORIAMENTO REMOTO/DIRPED/IBAMA

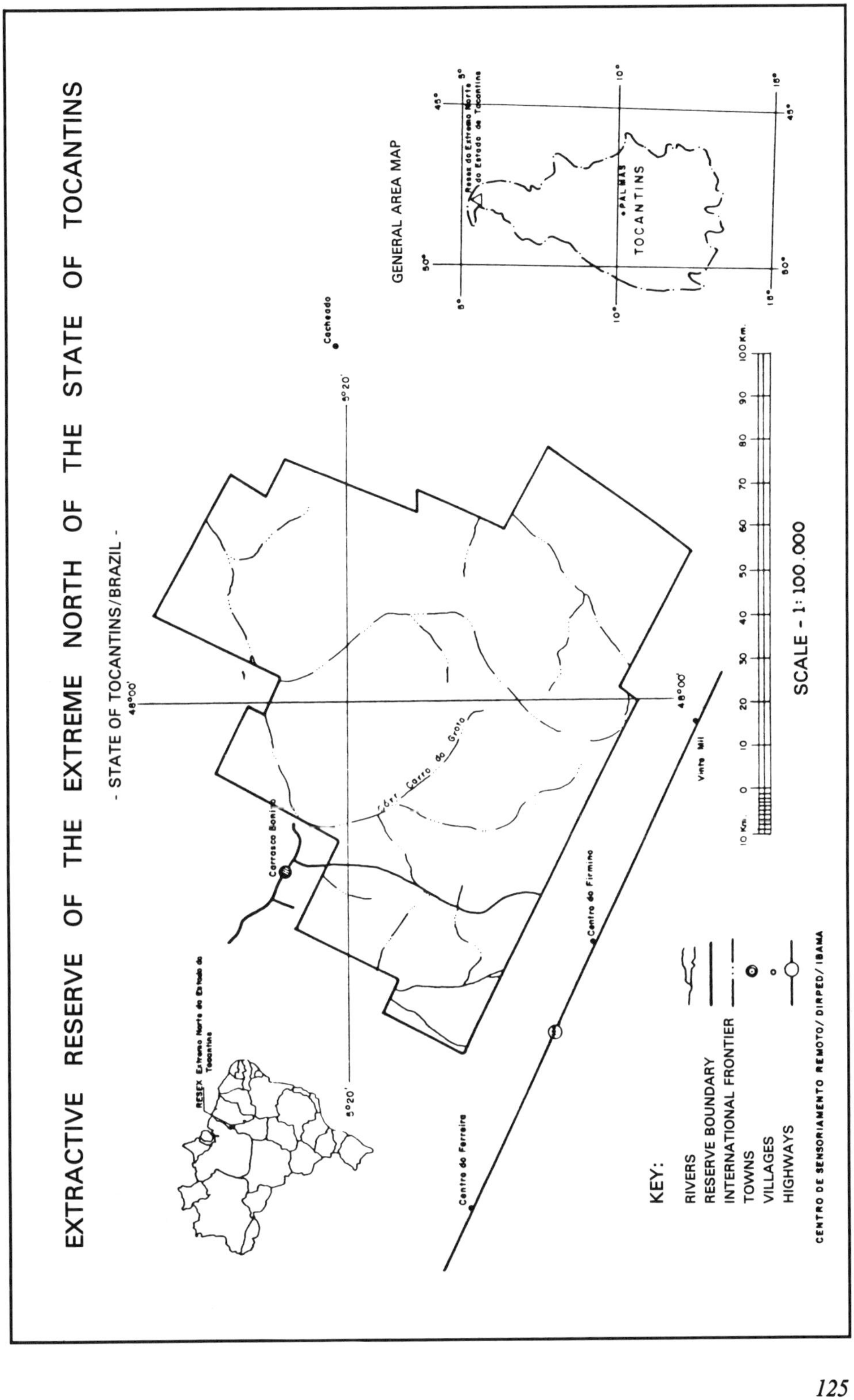
EXTRACTIVE RESERVE OF THE EXTREME NORTH OF THE STATE OF TOCANTINS
- STATE OF TOCANTINS/BRAZIL -
GENERAL AREA MAP
TOCANTINS
PALMAS
KEY:
RIVERS
RESERVE BOUNDARY
INTERNATIONAL FRONTIER
TOWNS
VILLAGES
HIGHWAYS
SCALE - 1: 100.000
Carrasco Bonito
Cacheado
Centro do Ferreira
Centro do Firmino
Vinte Mil
48°00'
5°20'
CENTRO DE SENSORIAMENTO REMOTO/ DIRPED/ IBAMA

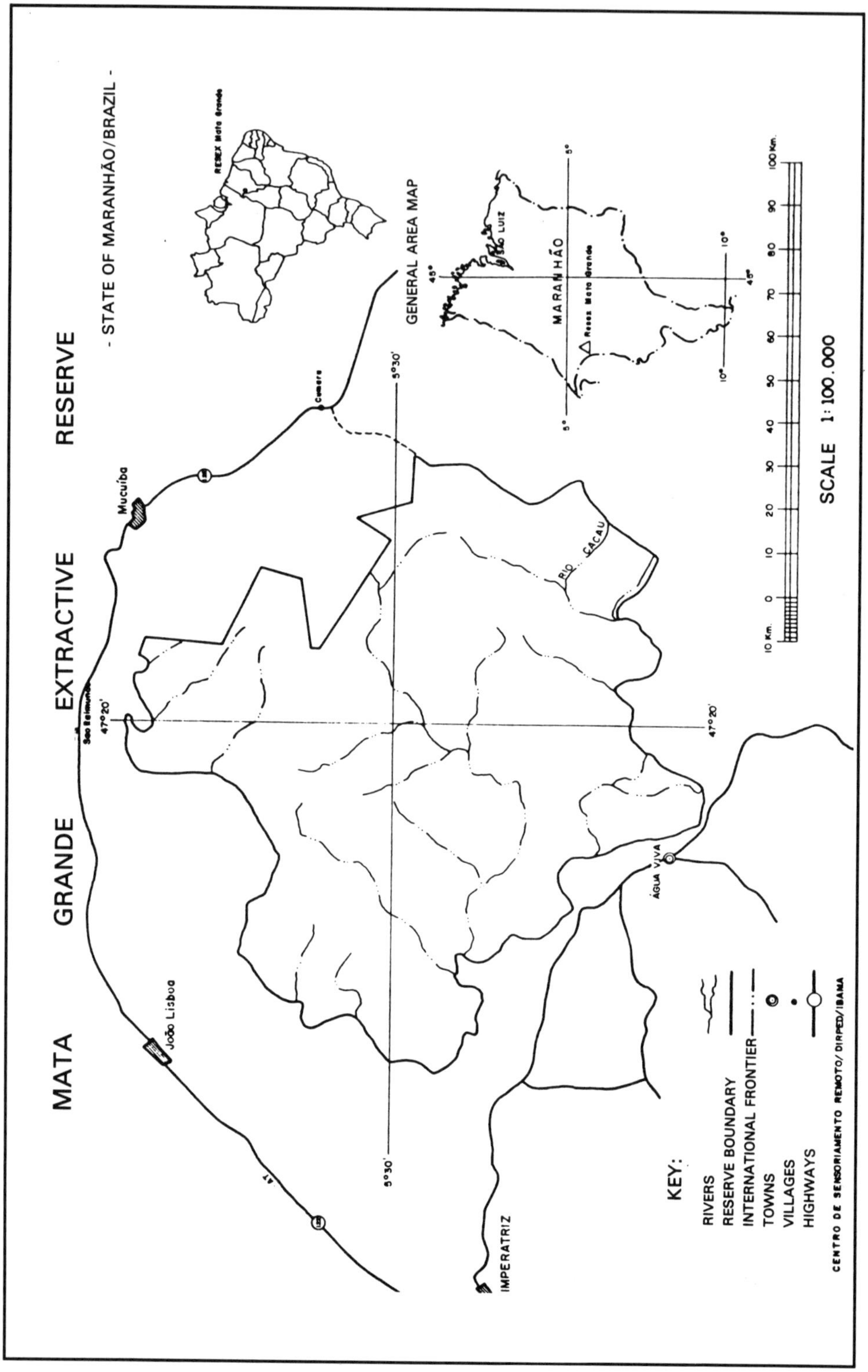
MATA GRANDE EXTRACTIVE RESERVE
- STATE OF MARANHÃO/BRAZIL -
GENERAL AREA MAP
MARANHÃO
SÃO LUIZ
SCALE 1:100.000
João Lisboa
Mucuiba
RIO CACAU
ÁGUA VIVA
IMPERATRIZ
5°30'
47°20'
KEY:
RIVERS
RESERVE BOUNDARY
INTERNATIONAL FRONTIER
TOWNS
VILLAGES
HIGHWAYS
CENTRO DE SENSORIAMENTO REMOTO/ DIRPED/ IBAMA

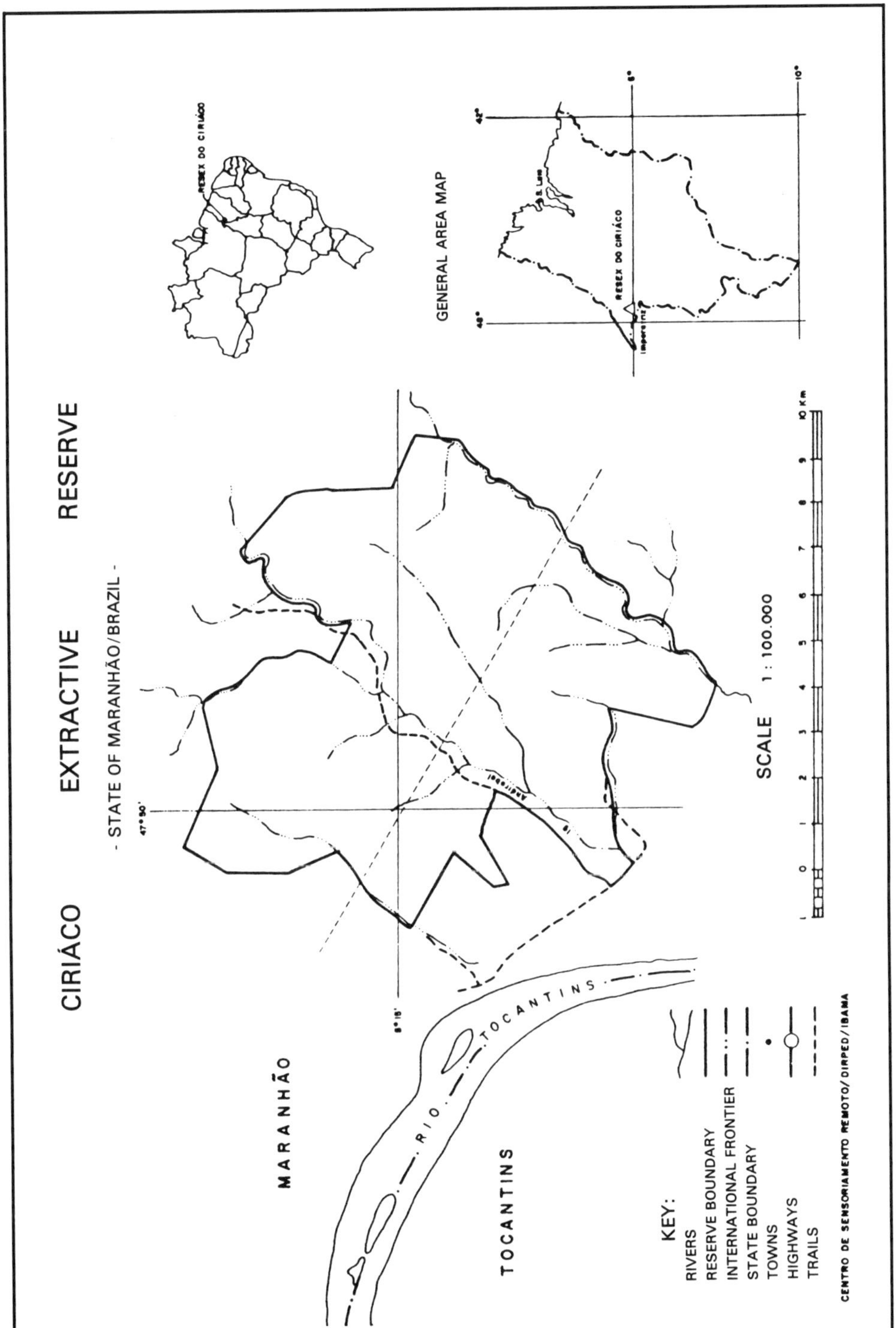
CIRIÁCO EXTRACTIVE RESERVE
- STATE OF MARANHÃO/BRAZIL -
47°30'
8°15'
MARANHÃO
TOCANTINS
RIO TOCANTINS
SCALE 1 : 100.000
1 0 1 2 3 4 5 6 7 8 9 10 Km
GENERAL AREA MAP
RESEX DO CIRIÁCO
Imperatriz
42°
48°
6°
10°
KEY:
RIVERS
RESERVE BOUNDARY
INTERNATIONAL FRONTIER
STATE BOUNDARY
TOWNS
HIGHWAYS
TRAILS
CENTRO DE SENSORIAMENTO REMOTO/DIRPED/IBAMA

BIBLIOGRAPHY

ALLEGRETTI, Mary Helena. Extrativismo e Ideologia. O Processo de Transformação do Seringueiro Cativo em Liberto. 1982 - Mimeografado.

ALLEGRETTI, Mary Helena. Os Seringueiros - Estudo de Caso em Seringal Nativo do Acre. Dissertação de Mestrado, UNB. Brasília, 1979.

ALLEGRETTI, Mary Helena. Reservas Extrativistas: Parâmetros para uma Política de Desenvolvimento Sustentável na Amazônia. In R.Bras Geog. Rio de Janeiro, 54 (1) 5-23 - Jan/Mar/92.

ANDERSON, Anthony B. Extrativismo Vegetal e Reservas Extrativistas: Limitações e Oportunidades. 1992 - Mimeografado.

ASSEMA, O Genocídio das Quebradeiras de Coco Babaçu, Maranhão, 1990.

ASAREAJ (ASSOCIAÇÃO DOS SERINGUEIROS E AGRICULTORES DA RESERVA EXTRATIVISTA DO ALTO JURUÁ). Plano de Ação Semestral - 1993/1994. Datilografado.

ASAREAJ (ASSOCIAÇÃO DOS SERINGUEIROS E AGRICULTORES DA RESERVA EXTRATIVISTA DO ALTO JURUÁ). História do Alto Juruá, 1993.

BELLIA, Vitor, Introdução à Economia do Meio Ambiente - Xerox 87p.

BELLIA, Vitor - BIOONE, Edison D- Rodovias Recursos Naturais e Meio Ambiente - EDUFF-DNER, 1993

BENATTI, José Heder, Manual de Plano de Utilização para as Reservas Extrativistas - Brasília, 1993 - Mimeografado.

BRASIL, Presidência da República. O Desafio do Desenvolvimento Sustentável. Relatório do Brasil para a Conferência das Nações Unidas sobre Meio Ambiente e Desenvolvimento - Brasília, 1991.

CEDEPLAR (Centro de Desenvolvimento e Planejamento Regional). Migrações Internas na Região Norte: O Caso do Acre - Belo Horizonte, 1971. Mimeografado.

CTA (Centro dos Trabalhadores da Amazônia)/CNPT/IBAMA/BANCO MUNDIAL. Educação, Saúde e Nutrição. Estudo de Pré-investimento para o Projeto Reservas Extrativistas - Programa Piloto para a Proteção das Florestas Tropicais do Brasil - PPG-7. Rio Branco, 1993.

CNPT (Centro Nacional de Desenvolvimento Sustentado das Populações Tradicionais). Relatório Anual de Atividades, 1993.

CNPT/IBAMA/PNUD. Documento de Assistência Preparatória. Projeto de Desenvolvimento Sustentado das Áreas de Reservas Extrativistas, Brasília, 1993.

CNPT/IBAMA/ECOTEC. Projeto Castanha - Beneficiamento descentralizado de Castanha do Pará - Estudo da Viabilidade do Rio Cajarí - Amapá, 1992. Mimeografafo

CNPT/IBAMA / BANCO MUNIDIAL / ECOTEC. Estudo Pré-investimento Projeto Reservas Extrativistas, 1994.

CNPT/IBAMA / IEA / BANCO MUNDIAL. Levantamento Sócio-econômico da Reserva

Extrativista Rio Ouro Preto - Projeto Reservas Extrativistas, Programa Piloto para a Proteção das Florestas Tropicais do Brasil -PPG-7, 1994.

CNPT/IBAMA / BANCO MUNDIAL / CNS. Levantamento Sócio-econômico da Reserva Extrativista Rio Cajarí - Projeto Reservas Extrativistas, Programa Piloto para a Proteção das Florestas Tropicias do Brasil - PPG-7, 1994.

CNPT/IBAMA / BANCO MUNDIAL / /ALMEIDA, Mauro. Levantamento Sócio-econômico da Reserva Extrativista do Alto Juruá - Projeto Reservas Extrativistas, Programa Piloto para a Proteção das Florestas Tropicais do Brasil -PPG-7, 1994.

CODETEC - Companhia de Desenvolvimento Tecnológico. Exploração Industrial de Produtos Naturais da Floresta Tropical Úmida. Campinas, 1990.

CONSELHO NACIONAL DOS SERINGUEIROS (CNS)/IBAMA/CNPT. Levantamento Sócio-econômico da Reserva Extrativista Chico Mendes, 1994. Datilografado.

CONSELHO NACIONAL DOS SERINGUEIROS (CNS). Levantamento Sócio-econômico das Reservas do Vale do Acre/Purus, 1992.

CONSELHO NACIONAL DOS SERINGUEIROS (CNS). Diretrizes para um Programa de Reservas Exstrativsitas na Amazônia. Rio Branco, 1993.

COOPERATIVA AGROEXTRITIVISTA DE XAPURÍ (CAEX). Custo-Benefício da Descentralização do Processamento de Castanha do Pará para uma Micro-usina - 1992. Mimeografado.

CUNHA, Manuela Carneiro da. BROWN, Keith. ALMEIDA, Mauro. Annual Report. Marh 1, 1993 to February 28, 1994. Project: Can Traditional Forest - Dwellers self manage conservation areas? A probing experiment in the Juruá Extrative reserve, Acre, Brazil.

CUNHA, Manuela Carneiro. BROWN, Keith. ALMEIDA, Mauro. É possível a Auto-Gestão de Áreas de Conservação por Populações Tradicionais? Uma Experiência Demonstrativa na Reserva Extrativsita do Alto Juruá/Acre - Brasil, 1993.

EMBRAPA - Proposta de uma Reserva Extrativista na Região do Rio Tejo, Campinas, 1989.

FUNTAC (Fundação de Tecnologia do Estado do Acre) - Monitoramento da Cobertura Florestal do Estado do ACre - Desmatamento e Uso da Terra - Rio Branco, 1990. Mimeografado.

HECHT, Susanna et COCKBURN, Alexander. The fatc of the Forest. London, 1989.

HOMMA, A.K. Uma Tentativa de Interpretação Teórica do Extrativismo Amazônico. In Acta Amazonica 12 (2);251-255, 1982. Mimeografado.

HOMMA, A.K. Reservas Extrativista: Uma Opção de Desenvolvimento Viável para a Amazônia? - In Para Desenvolvimento 25:38-48, 1981. Xerox.

IBAMA (Instituto Brasileiro do Meio Ambiente e dos Recursos Naturais Renováveis). Resumo de Informações. Proposta para criação da Reserva Extrativista do Ciriaco - Brasília, 1992.

IBAMA (Instituto Brasileiro do Meio Ambiente e dos Recursos Naturais Renováveis). Resumo de Informações. Proposta para criação da Reserva Extrativista da Mata Grande - Brasília, 1992.

IBAMA (Instituto Brasileiro do Meio Ambiente e dos Recursos Naturais Renováveis). Resumo de Informações. Proposta para criação da Reserva Extrativista do Extremo Norte do Tocantins - Brasília, 1992.

IBAMA (Instituto Brasileiro do Meio Ambiente e dos Recursos Naturais Renováveis). Resumo de Informações. Proposta para criação da Reserva Extrativista do Quilombo do Frexal - Brasília, 1992.

IBAMA/CNPT. Processo IBAMA No. 02001.001912/92-24. Propõe a criação da Reserva Extrativista do Ciriaco - Brasília, 1992.

IBAMA/CNPT. Processo IBAMA No. 02001.001910/92-07. Propõe a criação da Reserva Extrativista Mata Grande - Brasília, 1992.

IBAMA/CNPT. Processo IBAMA No. 02001.001913/92-87. Propõe a criação da Reserva Extrativista do Extremo Norte do Estado do Tocantins - Brasília, 1992.

IBAMA/CNPT. Processo IBAMA No. 02001.001911/92-61. Propõe a criação da Reserva Extrativista Marinha do Pirajubaé - Brasília, 1992.

IBAMA/CNPT. Processo IBAMA No. 02001.001927/92-00. Propõe a criação da Reserva Extrativista do Frechal - Brasília, 1992.

IBAMA/CNPT / BANCO MUNDIAL / IDEAS (Instituto de Desenvolvimento Ambiente e Sociedade). Levantamento Preliminar de Base de Recrusos Naturais - Reservas Extrativistas do Alto Juruá, Chico Mendes, Rio Ouro Preto e Rio Cajarí - Estudo Pré-investimento Projeto Reservas Extrativistas - 1993.

IBAMA/CNPT / BANCO MUNDIAL / IAC (Instituto de Assuntos Culturais). Fortalecimento Gerencial das Reservas Extrativistas - relatório, 1993 - Estudo de Pré-investimento para o Projeto Reservas Extrativistas - Programa Piloto para a Proteção das Florestas Tropicais do Brasil - PPG-7.

IBAMA/CNPT / IEA (Instituto de Estudos Amazônicos e Ambientais). Subsídios do IEA ao Projeto Reservas Extrativistas - Programa Piloto para a Proteção das Florestas Tropicais do Brasil - PPG-7, 1993.

IEA (Instituto de Estudos Amazônicos e Ambientais). Potencialidade dos Produtos Florestais não derivados da Madeira para Exploração Sustentável da Amazônia. Curitiba, 1991. Xerox.

Instituto Sociedade População e Natureza (ISPN) / IBAMA/CNPT / BANCO MUNDIAL. Projeto Reservas Extrativistas - Estudo de Impacto Ambiental, 1993.

MAY, Peter. Palmeiras em Chamas - Transformação Agrária e Justiça Social na Zona do Ba'bacu. EMAPA-FINEP - São Luis, 1990.

PINZÓN, Rueda Rafael. O Extrativismo - Mimeografado. 38p. Brasília, 1994.

Presidência da República - Secretaria do Meio Ambiente - SEMAM - Programa Emergencial para o Desenvolvimento Sustentado de Populações Tradicionais das Reservas Extrativistas - Brasília, 1992.

Projeto RADAMBRASIL - Ministério das Minas e Energias - Departamento Nacional de Produção Mineral - Vol. 13 - Levantamento de Recursos Naturais - Rio de Janeiro, 1977/

Projeto Seringueiro - CTA - Livro dos Seringueiros/1994.

Programa Nacional de Conservação e Desenvolvimento Florestal Sustentado - SEMAM/IBAMA - Brasília, 1991.

Programa Piloto para a Proteção das Florestas Tropicais do Brasil -Projeto Reservas Extrativistas - CNPT/IBAMA, 1994.

RODRIGUES, Écio. Mapeamento das Relações Sócio-econômicoas das Reservas Extrativistas Cachoeira e São Luis do Remanso. FUNTAC, Rio Branco, 1991.

SEMATUR/SAGRIMA/EMAPA. Workshop Babaçu sobre Alternativas Políticas, Sociais, e Tecnológicas para o Desenvolvimento Sustentável, Anais - São Luis, 1992.

SEMATUR. Diagnóstico Ambiental dos Principais Problemas Ambientais do Estado do Maranhão - São Luis, 1991.

SUDAM/BASA/SUFRAMA/PNUD. Cenários da Amazônia. 55p Mimeografado.

SUDAM. Plano de Desenvolvimento da Amazônia, 1992/95 - Belém, 1992.

SUDENE/Instituto Estadual do Babaçu. Mapeamento das Ocorrências e Prospecção do Potencial Atual do Babaçu no Estado do Maranhão - São Luis, 1981.

TORRES, Haroldo/MARTINE, George. Amazonian Extractivism: Prospects and Pitfalls - ISPN/Brasília, 1991 - Mimeografado.

UICN (1993) El Extractivismo en América Latina. Conclusiones, Recomendaciones del taller UICN-CEE. Amacayocu, Colombia, 1992.

WHITESELL, Edward Albert. Changing Courses: The Juruá River, its people and Amazonian Extractive Reserves. Dissetation University of California at Berkeley. 1993.